THE LITTLE ENGINEER COLORING BOOK

CARS AND TRUCKS

SETH MCKAY

The Little Engineer Coloring Book: Cars and Trucks by Seth McKay
www.TheLittleEngineerBooks.com

Copyright © 2018 by Seth McKay

All rights reserved. No portion of this book may be reproduced, stored in a retrieval system, or transmitted in any form or by any meanselectronic, mechanical, photocopy, recording, scanning, or other-except for brief quotations in critical reviews or articles, without the prior written permission of the publisher.

Creative Ideas Publishing titles may be purchased in bulk for educational, business, fund-raising, or sales promotional use. For information, please email permissions@TheLittleEngineerBooks.com.

ISBN-13: 978-1-952016-03-5

Published by: Creative Ideas Publishing

Table of Contents

Introduction (*for Parents*) v

Introduction (*for the Little Engineers*) vi

Tips on Using this Book vii

The First Cars 1

The First Trucks 2

Car Types 3

Where Cars are Made 4

Car Frame 5

Engine 6

Engine Cylinder 7

What a Gas Engine Needs 8

Diesel Engines 9

How Many Cylinders? 10

The Engine Block 11

Parts on the Engine Block 12

Engine Accessories 13

Radiator 14

Superchargers and Turbochargers 15

Superchargers 16

Turbochargers 17

Twin Turbochargers 18

Electric Motors 19

Supercars with Electric Motors 20

Front, Mid and Rear Engine 21
Transmission 22
Driveshaft 23
Differential 24
4-Wheel Drive 25
How 4WD Works 26
2 Differentials 27
4x4 28
Front Wheel Drive 29
Suspension 30
Solid Axle vs. Independent Suspension 31
Steering 32
Tires 33
Wheels 34
Brakes 35
Lights 36
Power Windows 37
Windshield Wipers 38
Inside the Car 39
Instrument Cluster 40
How Do Cars Get Fuel? 41
Electric Car Recharge 42
Certificate of Completion 43
BONUS: Special Preview of Space and Rockets Coloring Book 43

Introduction *(for parents)*

Thank you so much for your interest in The Little Engineer coloring books. In these books, your child will be introduced to new, interesting topics and begin to understand how things work.

I've always wanted my children to understand how things work, and in general, understand that objects are not magical boxes that work for some unknown reason. I want them to understand that objects are quite simple when broken down into a few key components. Without any direction, it is possible for a child to get the wrong understanding of objects that we consider to be simple.

For example, we understand that a printer has a roller that moves the paper and a little ink jet that sprays the paper with ink as it rolls by to make images. However, it would be very easy for our children to see the same printer and conclude that this magic box shoots out paper, and this is where paper comes from.

I chose to make this book a coloring book because I wanted to ensure that the book was also fun! I don't want this book to be too serious, boring or overwhelming. Coloring helps keep the book fun while incorporating new and challenging topics.

Book Objective:

The objective of this book is not to make your child an engineer (I hope your child chooses whatever career path suits him or her best), and the objective is not for your child to know exactly how everything works and memorize each part of a car.

The objective is simply to begin to introduce your child to new, interesting concepts that might catch his or her attention and help your child understand that seemingly complex things are just a few simple items put together.

Introduction *(for the Little Engineers)*

Hello Little Engineer!

My name is Seth, and I'm the Chief Engineer of this book. We will have so much fun learning about different parts of a car.

We will start with some of the first cars ever made and see what those look like. We will talk about all the main parts of a car and even super cool parts like turbochargers and new electric motors.

Enjoy coloring, ask your parents lots of questions, and most of all have fun!

Tips on Using this Book

- **Read the book to your child** – Before coloring the book, read it one time to help your child get a sense of the "big picture" before they focus on a single page to color.

- **Add your own explanations** – You know your child best so take the wording on a page and expand on it to help your child understand.

- **Relate the pages to your surroundings** – As you and your child walk around, point out different features on cars or ask them to help you find the items on a vehicle. (For example, ask them to show you the differential on a vehicle.)

- **Use your own car as an example** – Ask your child if they want to see specific parts on your car. This will help them get a sense of how these parts look "in the real world." Warning: If you've just driven the car, many parts will be too hot to touch so be careful and don't let your child touch hot car parts.

LET'S GET STARTED!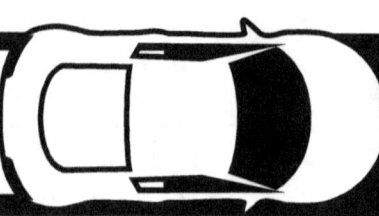

viii   The Little Engineer Coloring Book: Cars and Trucks

THE FIRST CARS

THE FIRST VEHICLE WAS MADE ALMOST 250 YEARS AGO! WOW, THAT IS A LONG TIME AGO. IT HAD 3 WHEELS AND A STEAM ENGINE, BUT NEVER DROVE VERY FAR. ABOUT 120 YEARS AGO, WAGONS WITH GASOLINE ENGINES WERE INVENTED.

THE FIRST TRUCKS

TEN YEARS AFTER THE FIRST GAS POWERED CAR, A TRUCK VERSION WAS MADE THAT WAS MORE POWERFUL AND COULD CARRY MORE CARGO.

CAR TYPES

COMPACT

SEDAN

TRUCK

SUV

VAN

SPORTS CAR

CARS AND TRUCKS COME IN MANY DIFFERENT SIZES.

HERE ARE SOME OF THE MAIN TYPES:
COMPACT, SEDAN, TRUCK, SUV, VAN, AND SPORTS CAR

WHERE CARS ARE MADE

MOST CARS ARE MADE IN LARGE FACTORIES AND THE FACTORY MAKES HUNDREDS OF CARS EVERYDAY.
THE CARS ARE THEN DRIVEN ONTO CAR TRANSPORT TRUCKS, TRAINS OR BOATS AND TAKEN ALL AROUND THE WORLD.

CAR FRAME

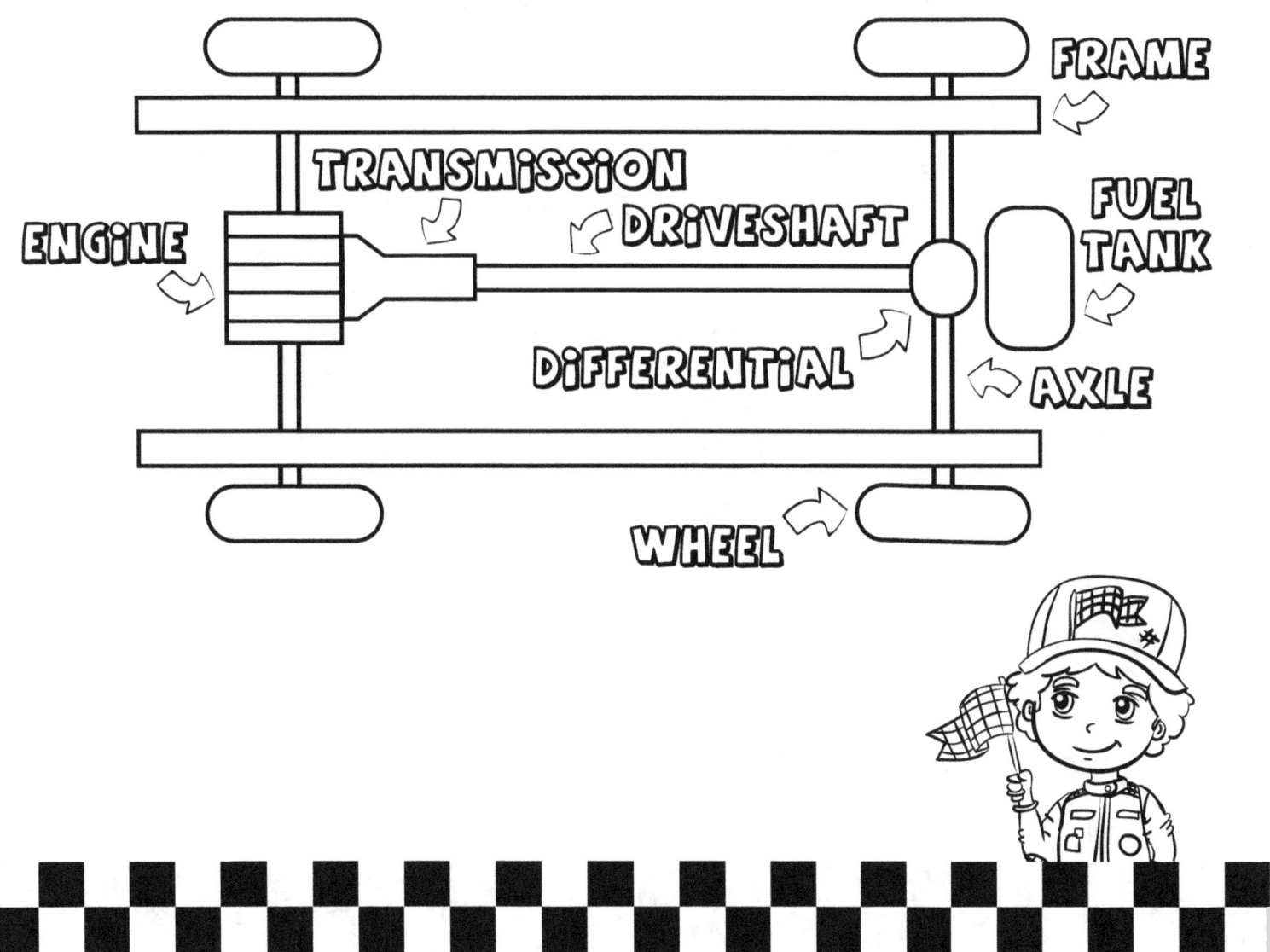

MOST CARS ARE BUILT ON A FRAME.

THE FRAME IS MADE OF METAL AND IS VERY STRONG SO THAT IT CAN HOLD ALL THE HEAVY PARTS OF THE CAR LIKE THE ENGINE, TRANSMISSION, SUSPENSION AND THE BODY OF THE CAR.

ENGINE

THE ENGINE IS ONE OF THE MOST IMPORTANT PARTS OF A CAR. THE ENGINE HAS MOVING PARTS THAT MAKE THE WHEELS SPIN SO THE CAR MOVES. SOME ENGINES ARE POWERED BY DIESEL FUEL WHILE OTHERS USE GAS (ALSO CALLED GASOLINE). SOME CARS HAVE ELECTRIC MOTORS INSTEAD OF GAS OR DIESEL ENGINES.

ENGINE CYLINDER

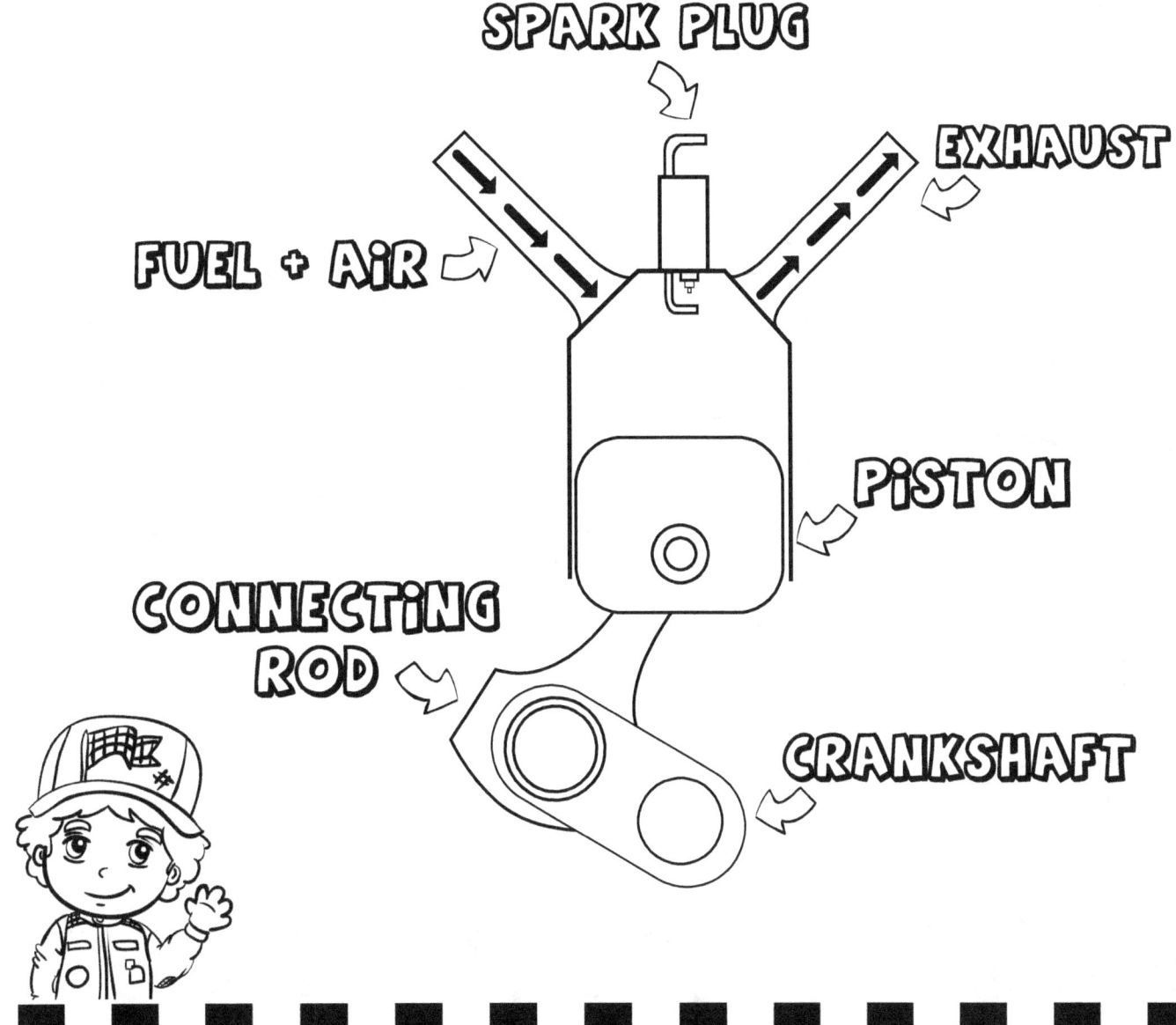

AN ENGINE HAS 1 OR MORE CYLINDERS IN IT. EACH CYLINDER HAS A PISTON INSIDE OF IT THAT MOVES UP AND DOWN WHICH MAKES THE CRANKSHAFT SPIN. THE CRANKSHAFT IS CONNECTED TO THE WHEELS, AND MAKES THE WHEELS SPIN TOO. THE PISTON MOVES BECAUSE FUEL GOES INTO THE CYLINDER AND MAKES A SMALL EXPLOSION WHEN THE SPARK PLUG LIGHTS THE FUEL ON FIRE.

The Little Engineer Coloring Book: Cars and Trucks

WHAT A GAS ENGINE NEEDS

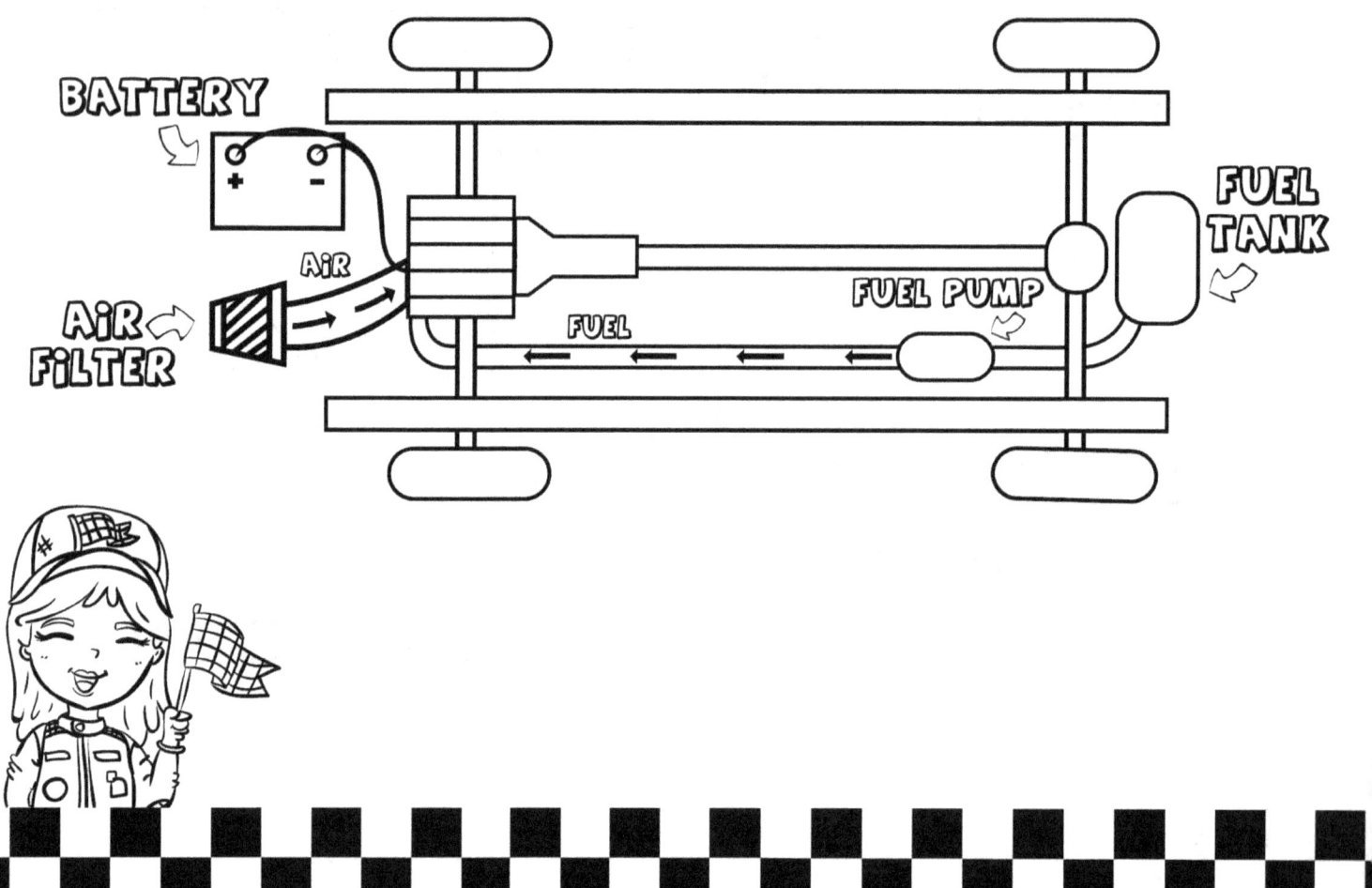

A GAS ENGINE NEEDS 3 THINGS TO WORK:
AIR, FUEL AND A SPARK.

- AIR GOES THROUGH THE AIR FILTER AND INTO THE ENGINE.
- A FUEL PUMP PUSHES THE GAS FROM THE GAS TANK TO THE ENGINE.
- SPARK PLUGS MAKE A SPARK. THE SPARK PLUGS NEED POWER SO THEY ARE CONNECTED TO THE BATTERY FOR ELECTRICITY.

DIESEL ENGINES

DIESEL ENGINES ONLY NEED AIR AND FUEL TO OPERATE. A DIESEL ENGINE SQUEEZES THE AIR AND FUEL SO TIGHT IN THE CYLINDER THAT THE FUEL EXPLODES BY ITSELF! NO SPARK PLUG NEEDED.

MOST BIG TRUCKS HAVE DIESEL ENGINES BECAUSE THE ENGINES ARE VERY POWERFUL AND CAN HELP CARRY HEAVY LOADS.

QUESTION TIME!

A GAS V8 ENGINE HAS 8 SPARK PLUGS. HOW MANY SPARK PLUGS DOES A DIESEL V8 ENGINE HAVE?

HOW MANY CYLINDERS?

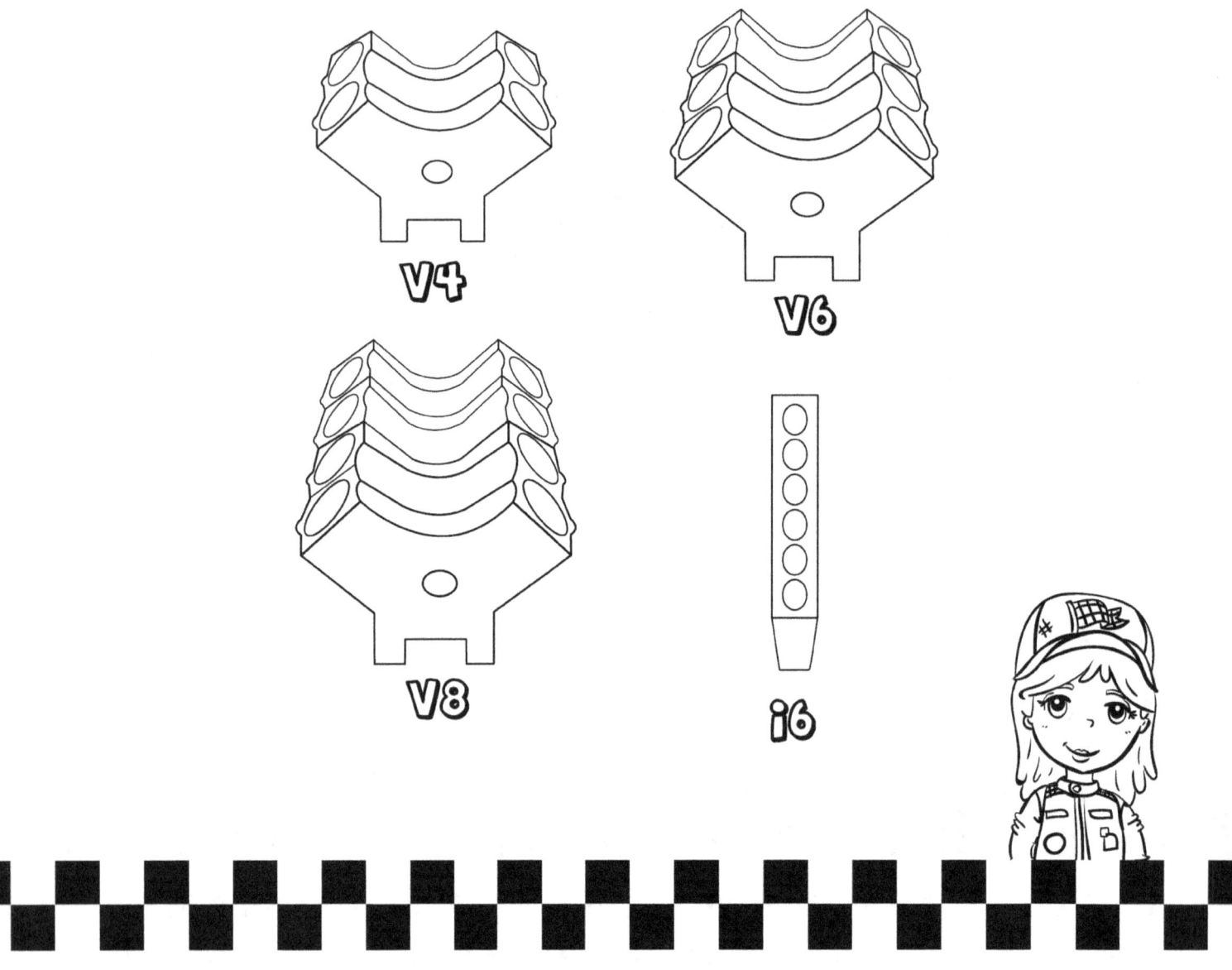

ENGINE NAMES ARE USUALLY BASED ON HOW MANY CYLINDERS THE ENGINE HAS.

A V8 ENGINE HAS 8 CYLINDERS.
AN I6 ENGINE HAS 6 CYLINDERS. IT IS ALSO CALLED AN INLINE 6 BECAUSE ALL THE CYLINDERS ARE IN A LINE.

QUESTION TIME!
HOW MANY CYLINDERS DOES A V6 ENGINE HAVE?

THE ENGINE BLOCK

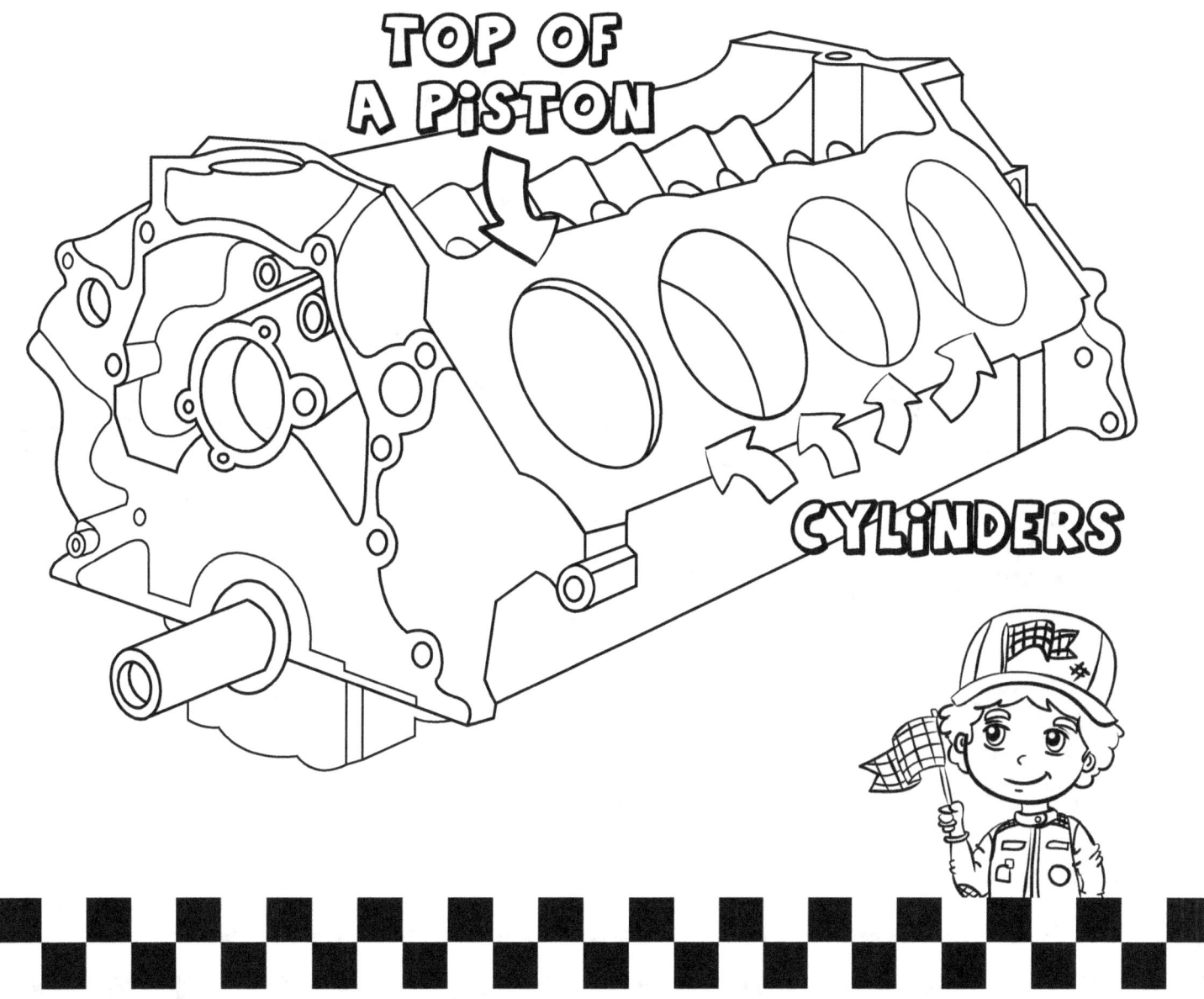

THE ENGINE BLOCK IS A BIG METAL PART WITH "HOLES" IN IT FOR EACH CYLINDER. THE PISTONS, CONNECTING RODS AND CRANKSHAFT ARE HIDDEN INSIDE OF THE ENGINE.

QUESTION TIME!
HOW MANY CYLINDERS CAN YOU SEE?

PARTS ON THE ENGINE BLOCK

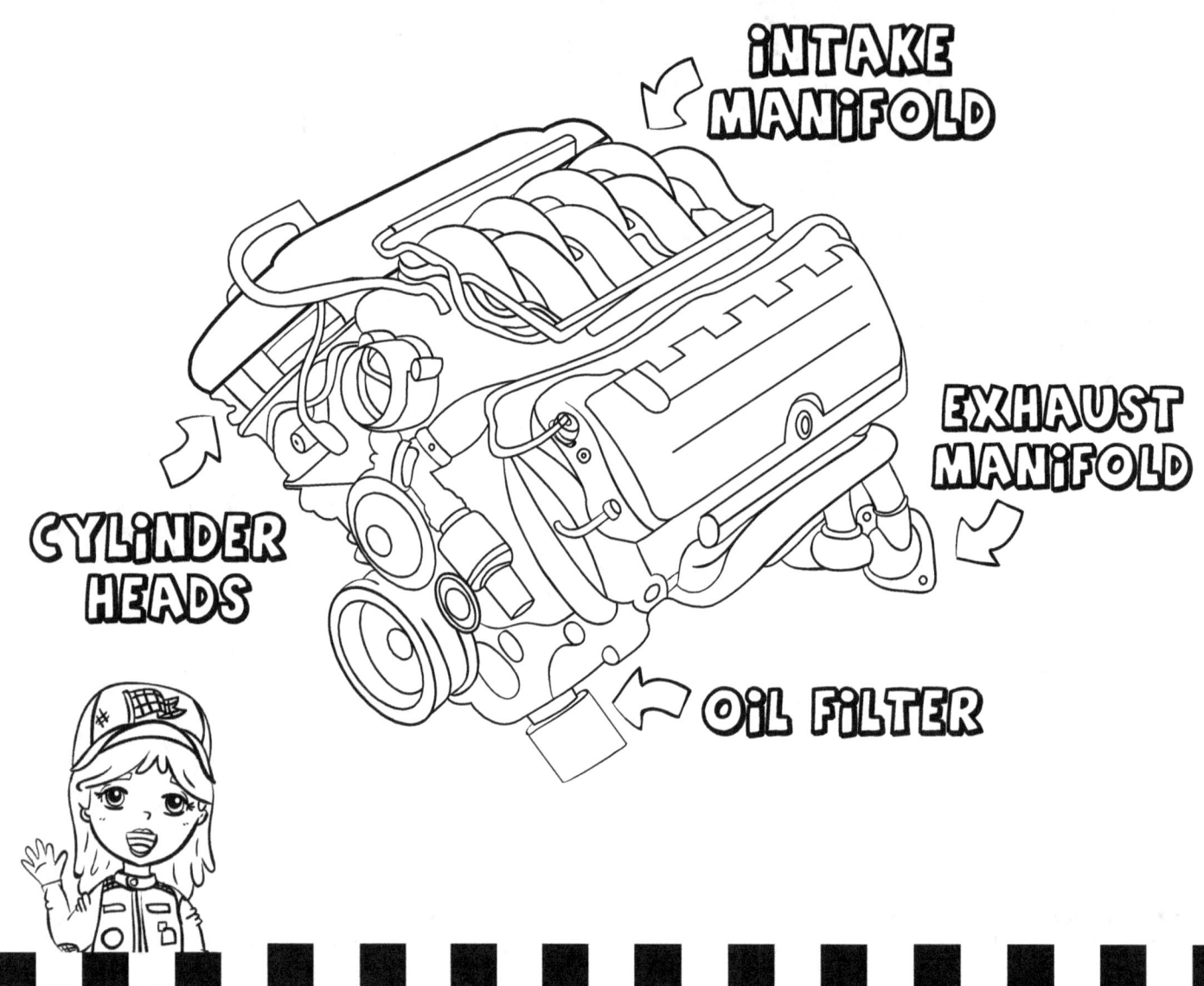

MANY PARTS GO ON TOP OF THE ENGINE BLOCK. CYLINDER HEADS COVER UP THE CYLINDER "HOLES." AN AIR INTAKE MANIFOLD IS ADDED ON TOP OF THE ENGINE. ALL THE AIR THAT GOES INTO THE CYLINDERS GOES THROUGH THE INTAKE MANIFOLD FIRST. THE EXHAUST MANIFOLDS CONNECT THE ENGINE TO THE EXHAUST PIPES THAT GO UNDER THE CAR.

ENGINE ACCESSORIES

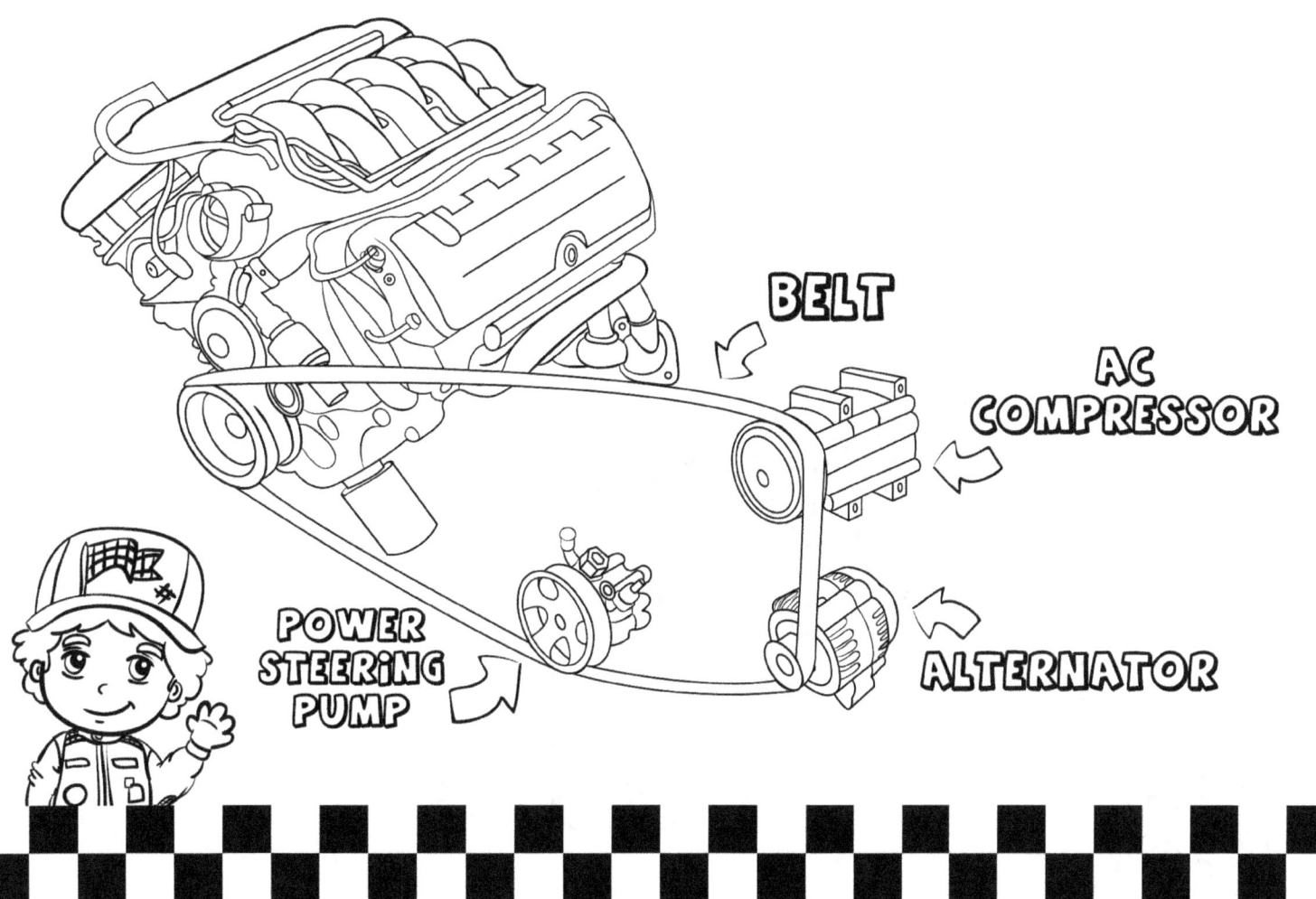

THE ENGINE ACCESSORIES ARE POWERED BY A BELT ON THE ENGINE. THE BELT SPINS THE WHEELS ON THE ACCESSORIES WHEN THE ENGINE IS ON.

- THE ALTERNATOR IS THE BATTERY CHARGER FOR THE CAR. IT IS A SMALL POWER GENERATOR THAT MAKES SURE YOUR CAR HAS PLENTY OF ELECTRICITY FOR LIGHTS, SPARK PLUGS, THE RADIO AND MORE.
- THE AC COMPRESSOR HELPS THE AC SYSTEM WORK SO THE AIR IS NICE AND COLD INSIDE THE CAR.
- THE POWER STEERING PUMP MAKES IT EASIER TO TURN THE STEERING WHEEL.

RADIATOR

ENGINES GET VERY HOT.

- A RADIATOR HELPS KEEP THE ENGINE FROM GETTING TOO HOT.
- A WATER PUMP MOVES A SPECIAL LIQUID CALLED COOLANT THROUGH THE ENGINE AND THEN THROUGH A RADIATOR.

SUPERCHARGERS AND TURBOCHARGERS

SUPERCHARGERS AND TURBOCHARGERS ACT LIKE A FAN THAT PUSHES EXTRA AIR INTO THE ENGINE.

THE EXTRA AIR ALLOWS THE ENGINE TO HAVE MORE POWER SO THE ENGINE CAN MAKE THE CAR GO EVEN FASTER!

SUPERCHARGERS

SUPERCHARGERS ARE FANS THAT ARE POWERED BY A BELT THAT SPINS WHEN THE ENGINE IS ON.

MOST SUPERCHARGERS ARE ON TOP OF THE ENGINE AND SOMETIMES IT IS SO BIG THAT YOU CAN SEE IT STICKING OUT OF THE HOOD!

TURBOCHARGERS

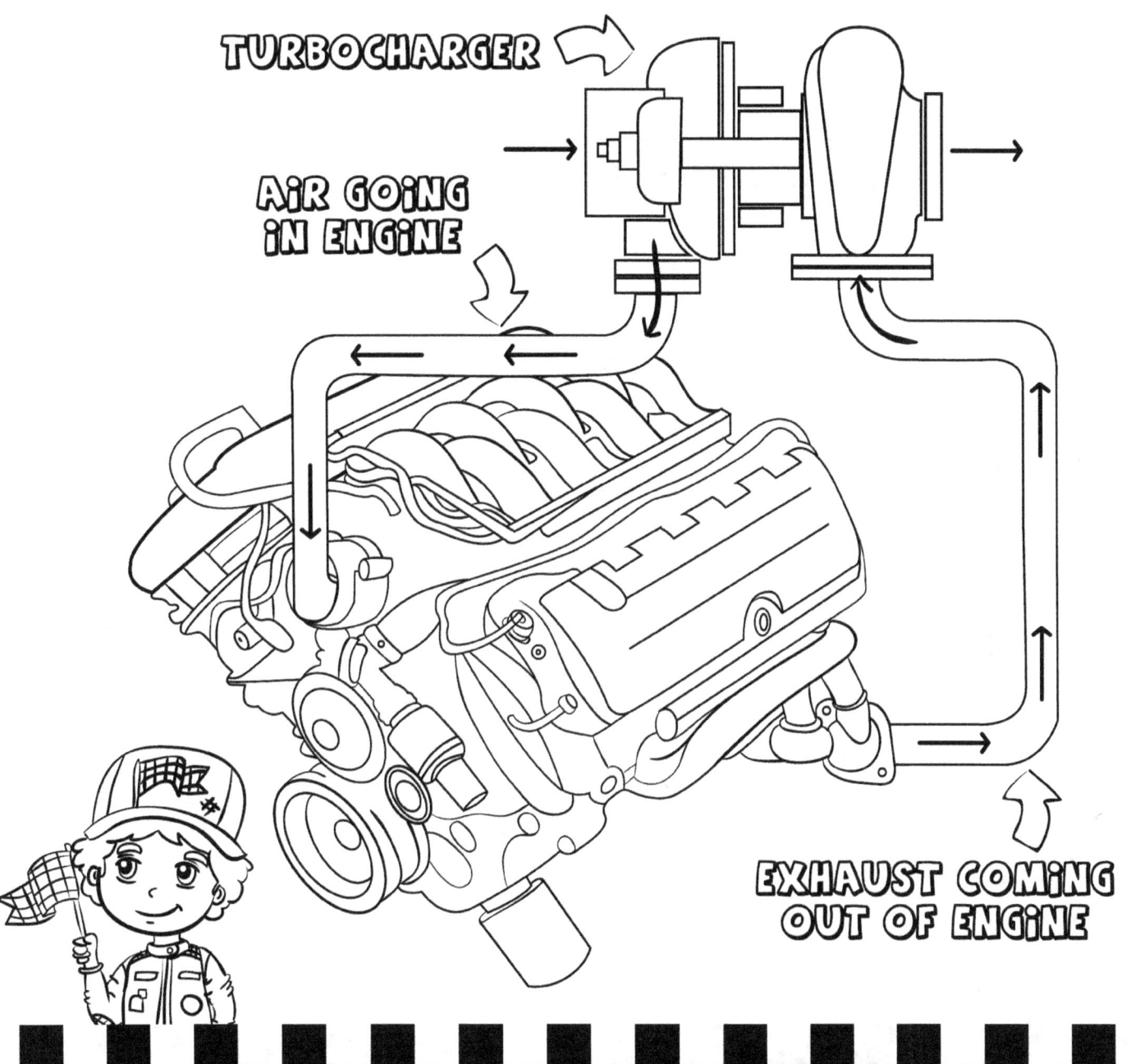

A TURBOCHARGER IS NOT POWERED BY A BELT. INSTEAD, IT IS LIKE A 2 SIDED FAN. THE EXHAUST SHOOTING OUT OF THE ENGINE SPINS ONE SIDE OF THE FAN SO THIS SPINS THE OTHER SIDE OF THE FAN WHICH PUSHES AIR INTO THE ENGINE.

TWIN TURBOCHARGERS

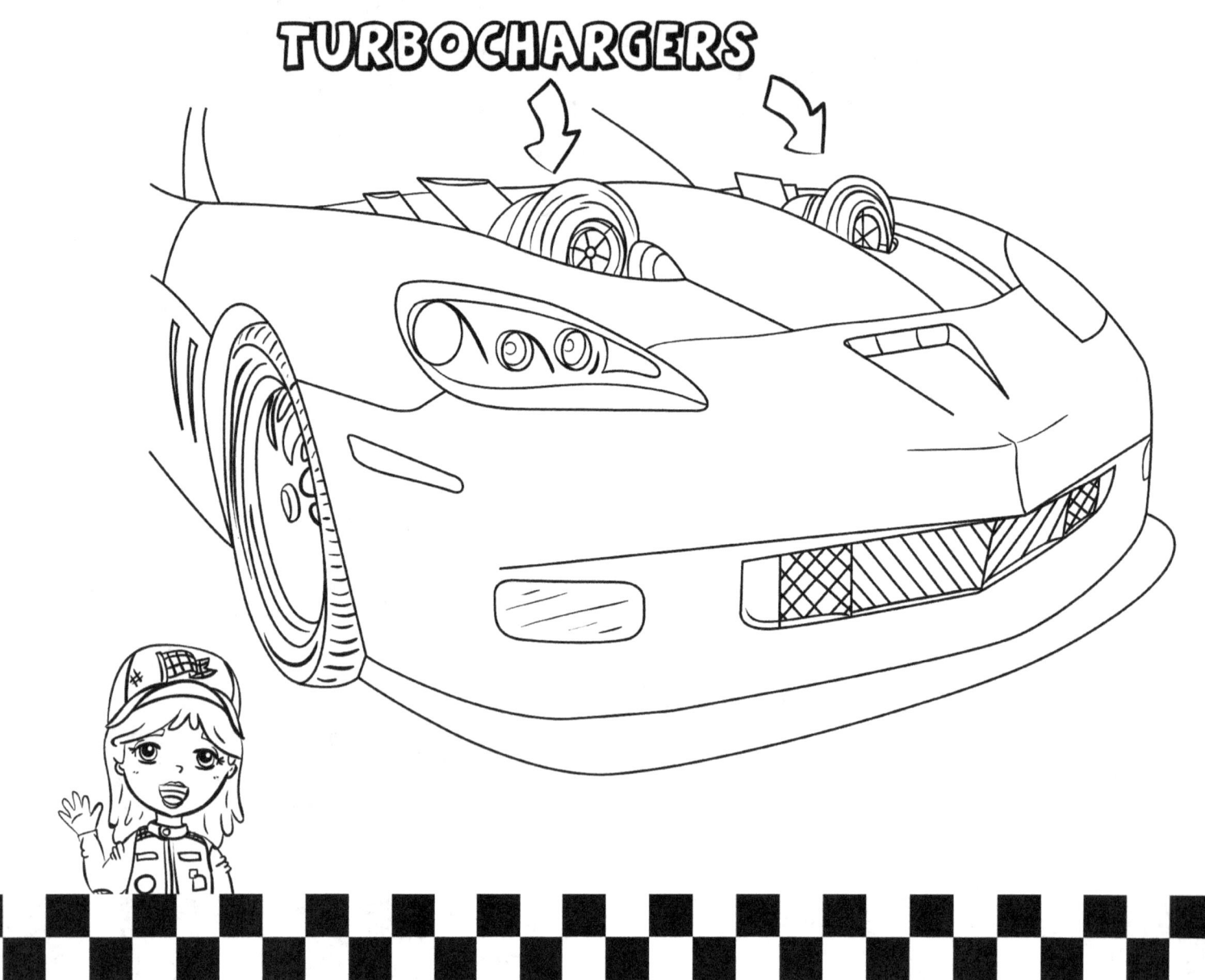

THIS CAR HAS 2 TURBOCHARGERS!

MOST TURBOCHARGERS ARE UNDER THE HOOD AND HARD TO SEE, BUT SOME CARS HAVE THEM STICKING OUT OF THE HOOD WHICH IS REALLY COOL!

ELECTRIC MOTORS

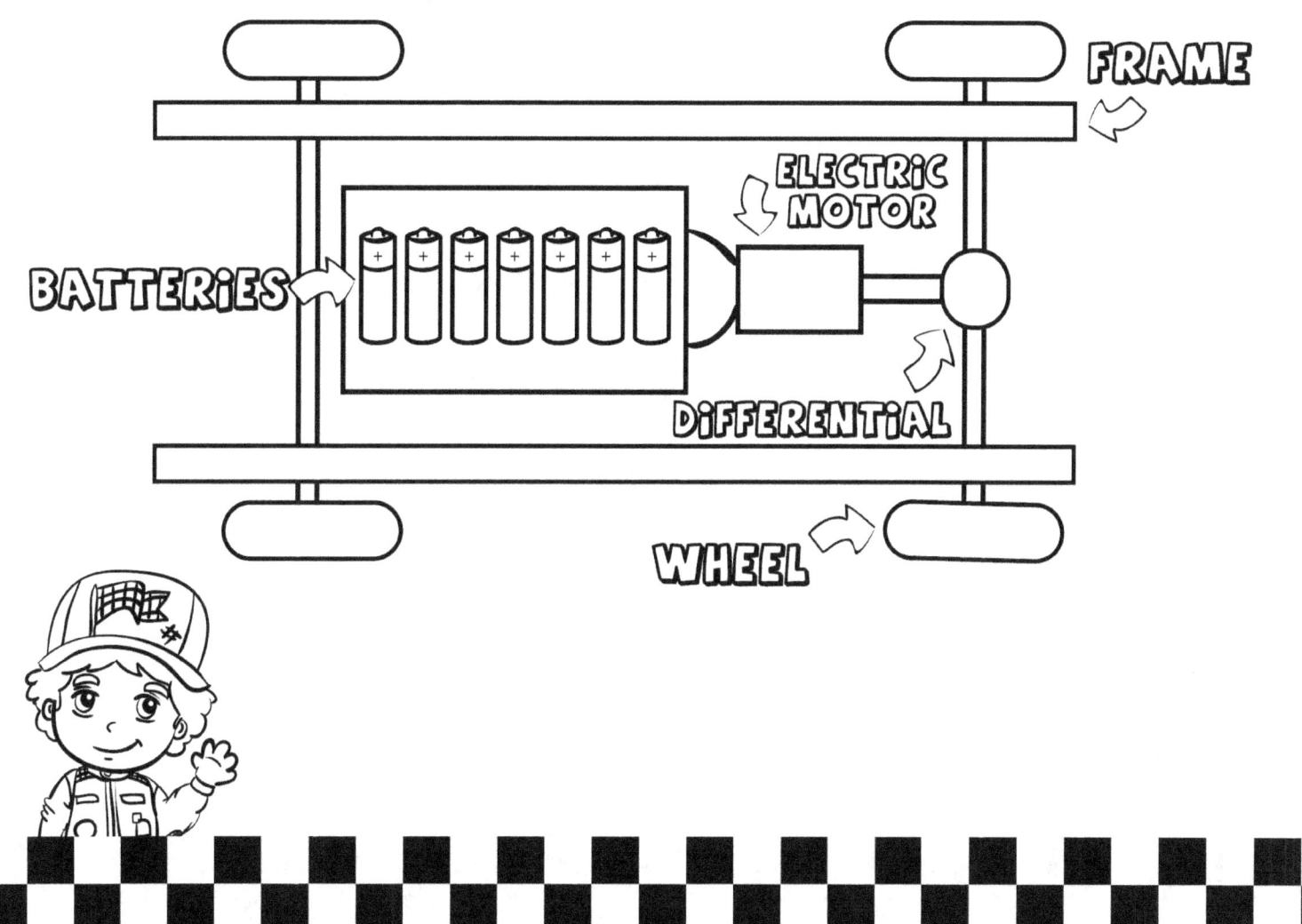

CARS ALSO USE ELECTRIC MOTORS. ELECTRIC MOTORS HAVE LESS MOVING PARTS WHICH HELPS THEM LAST LONGER.

THE MOTORS IN AN ELECTRIC CAR ARE POWERED BY BATTERIES. MOST REMOTE CONTROL CARS WORK THE SAME WAY!

SUPERCARS WITH ELECTRIC MOTORS

SOME OF THE FASTEST SUPERCARS IN THE WORLD HAVE ELECTRIC MOTORS!

FRONT, MID AND REAR ENGINE

THERE ARE FRONT, MID AND REAR ENGINE CARS.

- FRONT ENGINE CARS HAVE THE ENGINE BETWEEN THE FRONT WHEELS AND THE FRONT OF THE CAR.
- MID ENGINE CARS HAVE THE ENGINE BETWEEN THE FRONT AND REAR WHEELS.
- REAR ENGINE CARS HAVE THE ENGINE BETWEEN THE BACK WHEELS AND BACK OF THE CAR.

TRANSMISSION

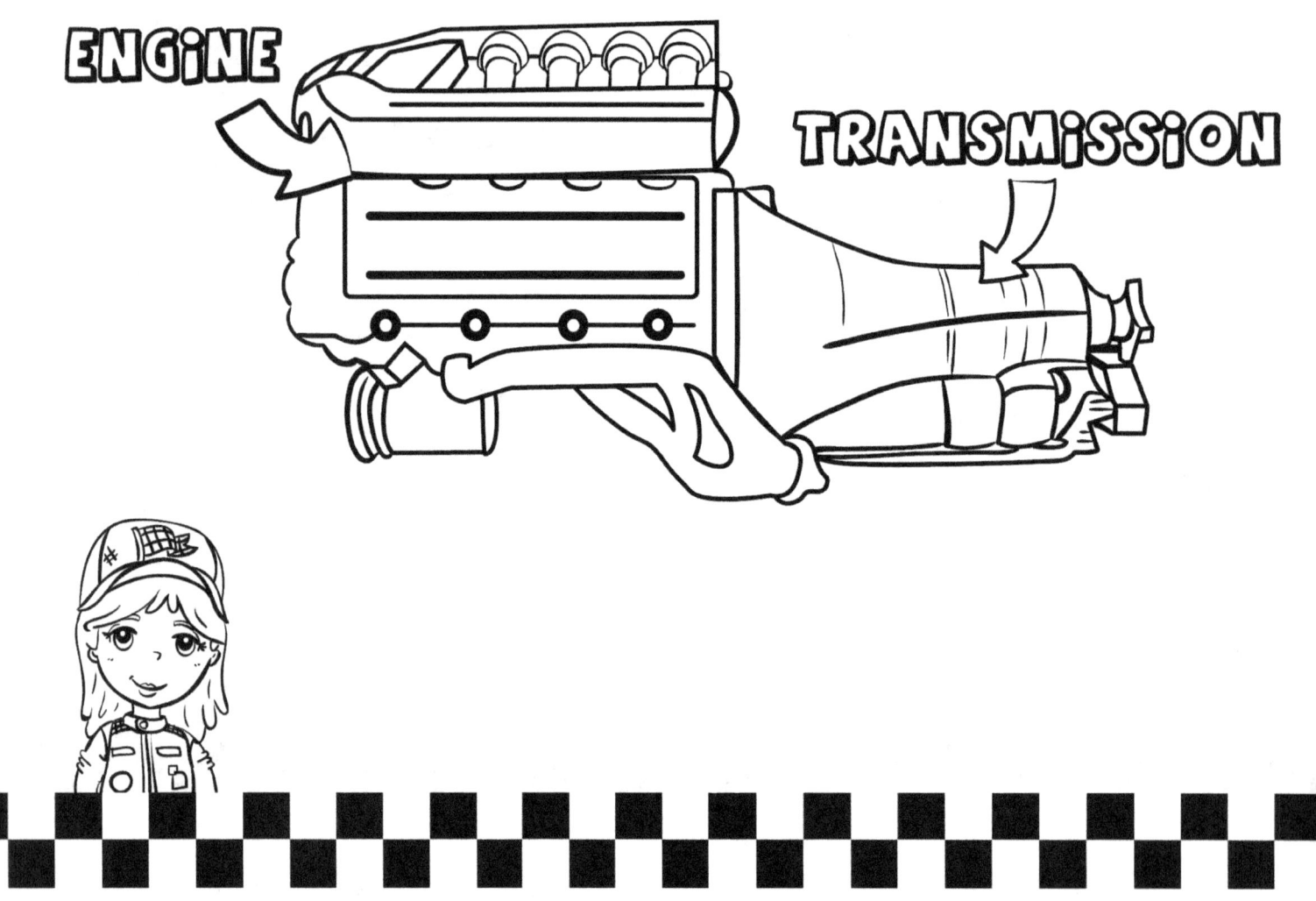

THE TRANSMISSION CONNECTS TO THE BACK OF THE ENGINE.

THE TRANSMISSION HAS DIFFERENT GEARS THAT HELP THE WHEELS SPIN AT THE RIGHT SPEED. THE TRANSMISSION ALSO HAS A GEAR CALLED 'REVERSE' SO THE CAR CAN MOVE BACKWARDS.

DRIVESHAFT

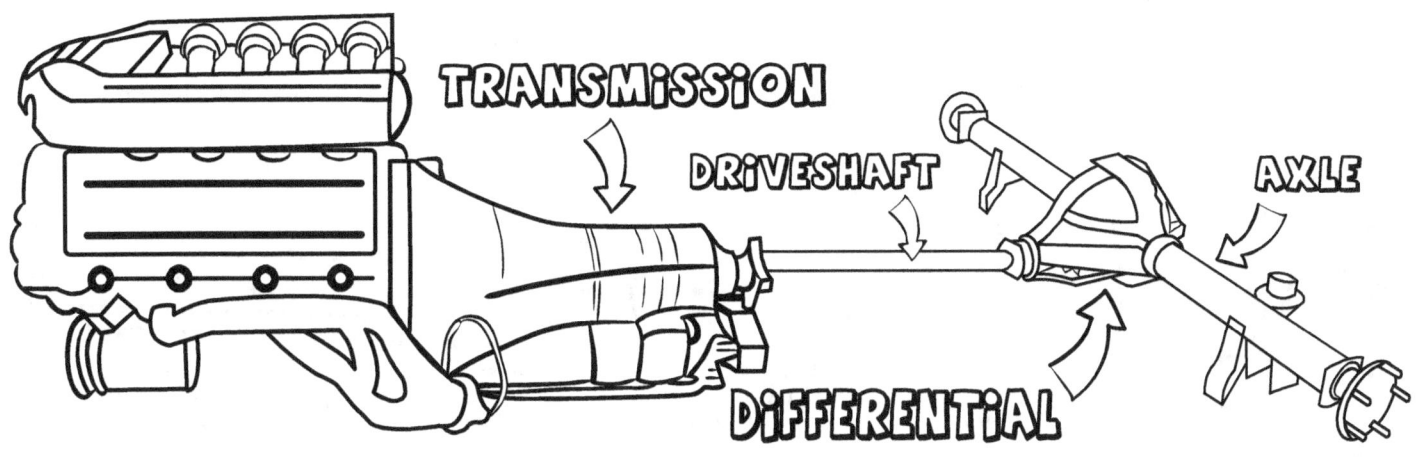

THE BACK OF THE TRANSMISSION IS CONNECTED TO A DRIVESHAFT WHICH IS CONNECTED TO THE DIFFERENTIAL. THE DIFFERENTIAL AND AXLE ARE CONNECTED TO THE WHEELS SO THAT THE WHEELS SPIN WHEN THE DRIVESHAFT SPINS.

DIFFERENTIAL

THE DIFFERENTIAL NEEDS TO BE VERY STRONG TO HANDLE ALL THE ENGINE'S POWER. YOU CAN SEE THE DIFFERENTIAL BETWEEN THE BACK WHEELS OF MOST TRUCKS. IT IS THE ROUND PART IN THE MIDDLE OF THE AXLE.

4-WHEEL DRIVE

IN MANY VEHICLES, ONLY THE BACK WHEELS SPIN WHICH IS CALLED "2-WHEEL DRIVE." SOME CARS AND TRUCKS HAVE "4-WHEEL DRIVE" WHICH MEANS THAT ALL 4 WHEELS ARE CONNECTED TO THE ENGINE. THIS ALLOWS THE VEHICLE TO DRIVE IN MUD OR OVER BIG ROCKS AND NOT GET STUCK.

HOW 4WD WORKS

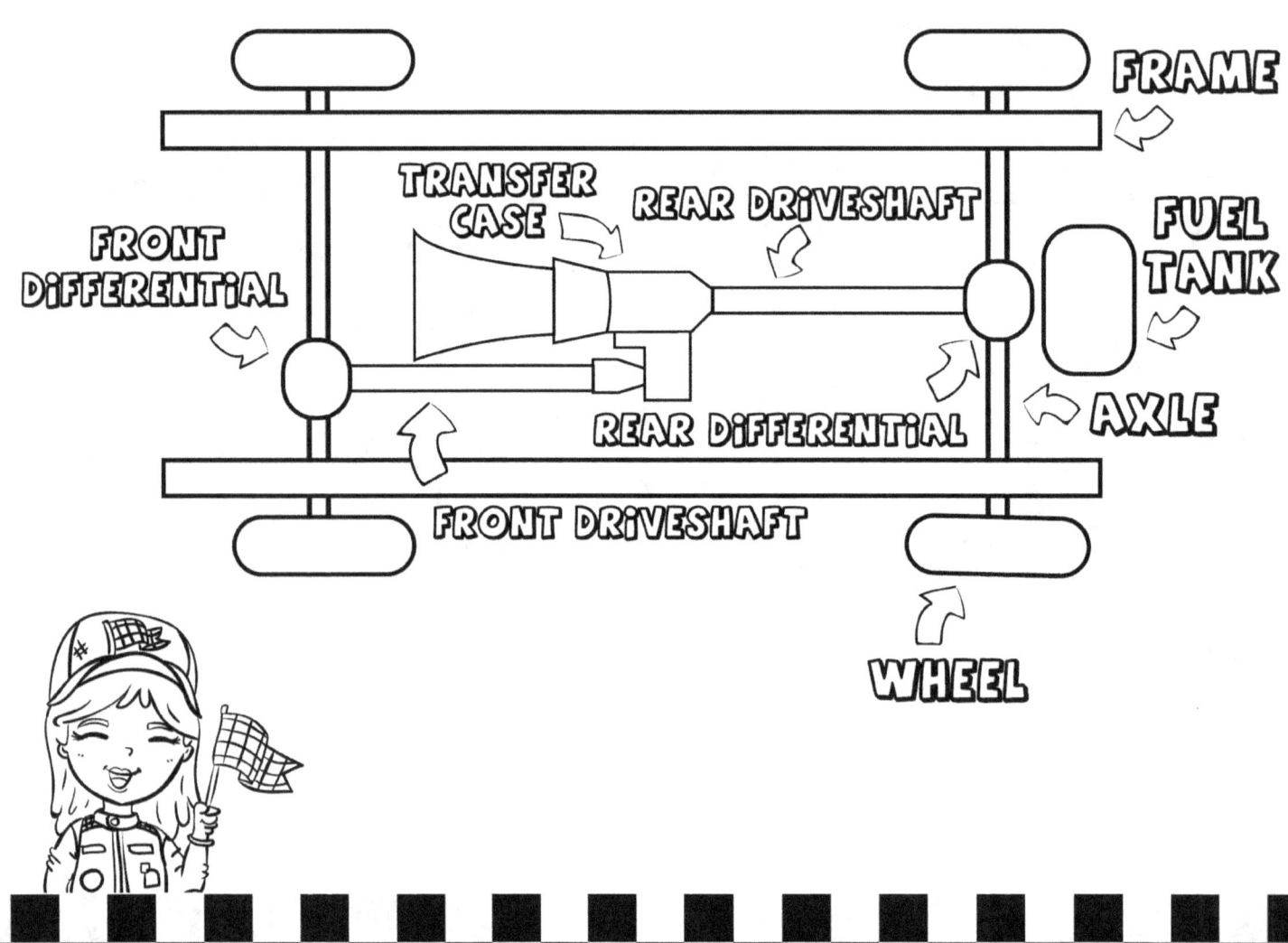

IN A 4-WHEEL DRIVE VEHICLE, A TRANSFER CASE IS ON THE BACK OF THE TRANSMISSION AND THIS CONNECTS TO THE FRONT AND REAL AXLES. THIS ALLOWS THE ENGINE TO SPIN THE FRONT AND BACK WHEELS.

2 Differentials

ON TRUCKS, YOU CAN SOMETIMES SEE 2 DIFFERENTIALS.
THIS MEANS THE TRUCK HAS 4-WHEEL DRIVE.
SOME CARS HAVE 4-WHEEL DRIVE, BUT IT IS HARD
TO SEE THE DIFFERENTIAL BECAUSE THE CAR IS
CLOSE TO THE GROUND.

4X4

4X4 MEANS THE SAME THING AS 4-WHEEL DRIVE. IF YOU CAN'T SEE THE FRONT DIFFERENTIAL, YOU CAN LOOK FOR THE "4X4" SYMBOL TO SEE IF A TRUCK IS 4-WHEEL DRIVE.

FRONT WHEEL DRIVE

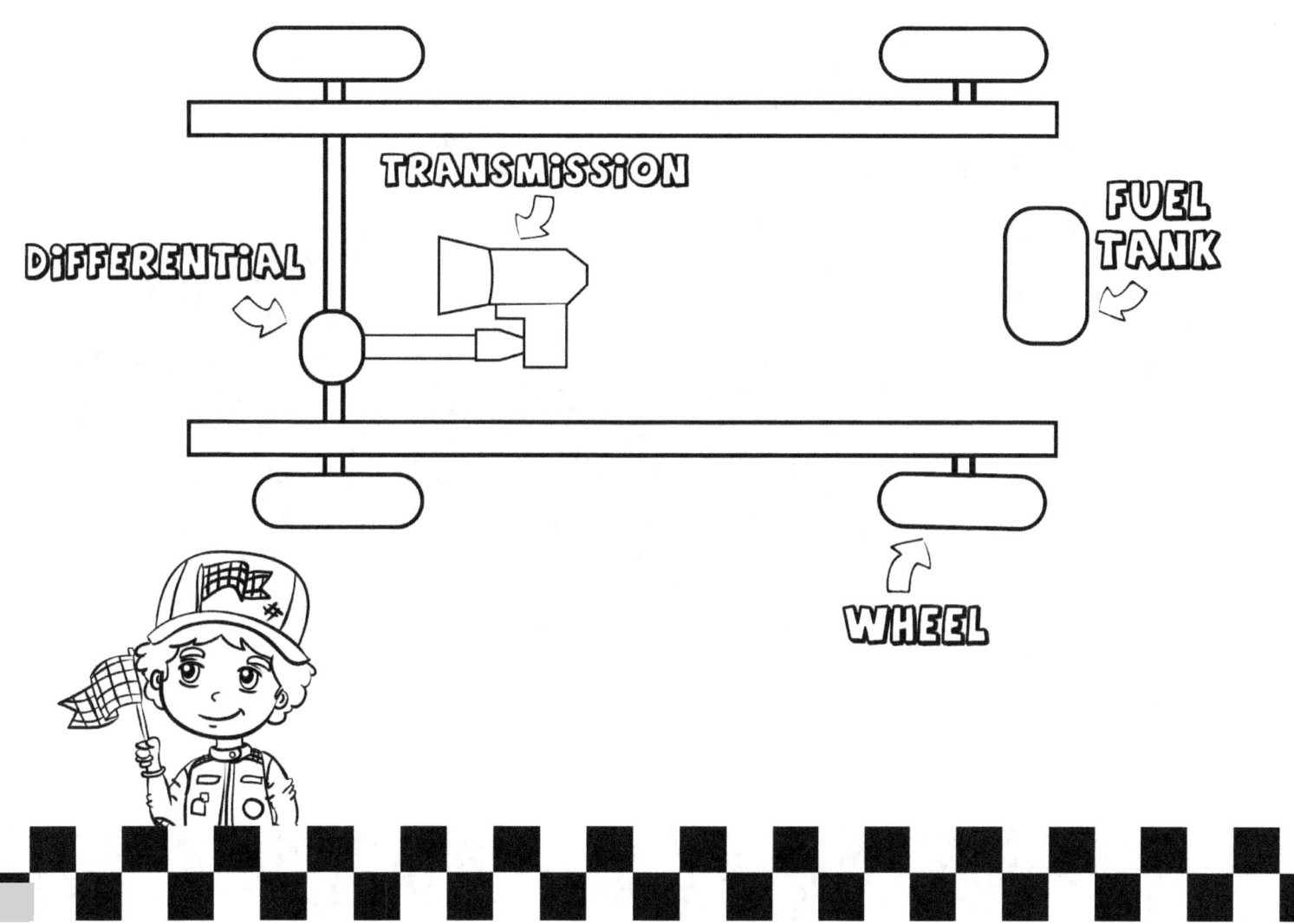

SOME 2-WHEEL DRIVE CARS ONLY POWER THE FRONT WHEELS; THIS IS CALLED FRONT WHEEL DRIVE.

The Little Engineer Coloring Book: Cars and Trucks

SUSPENSION

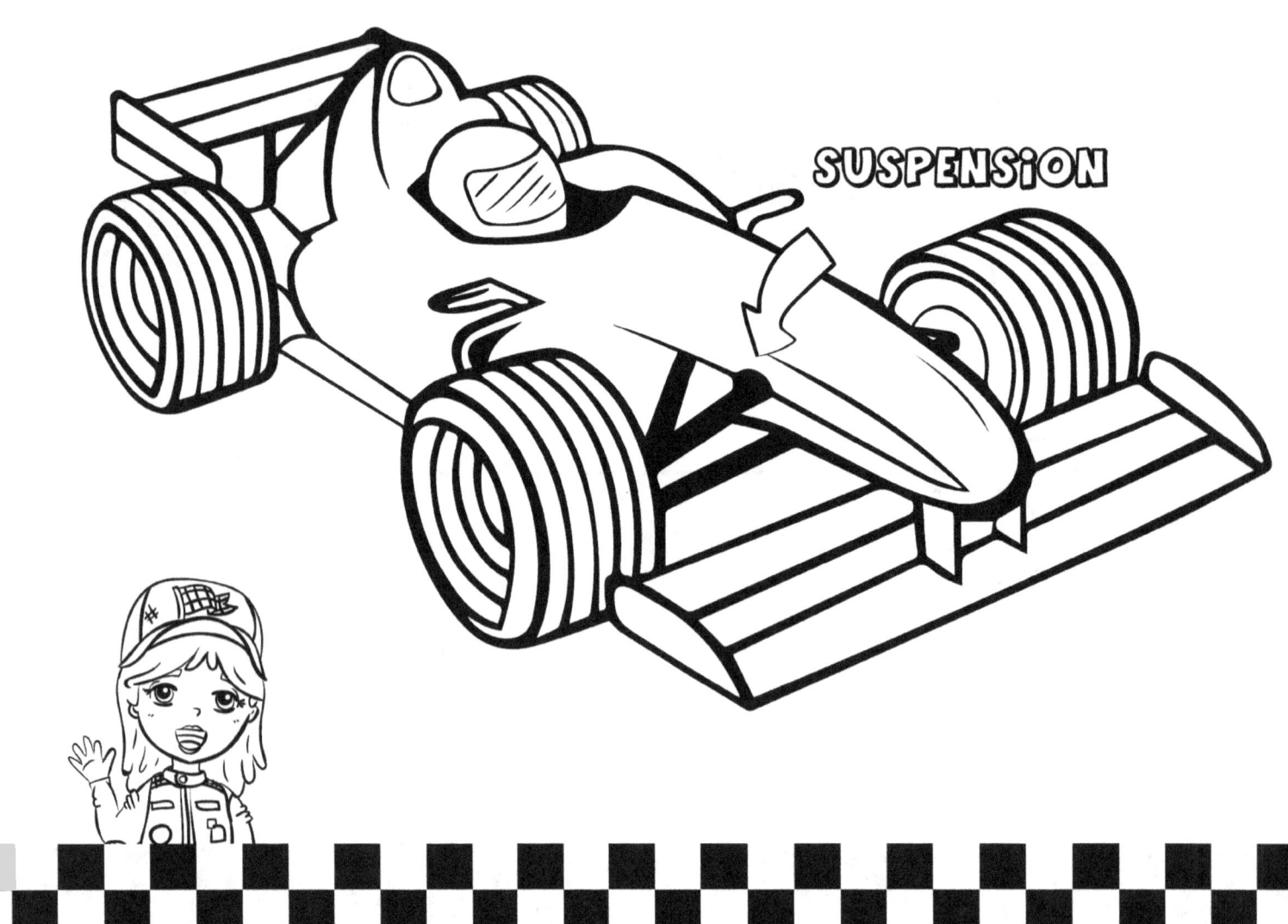

SUSPENSION CONNECTS THE WHEELS TO THE FRAME OR BODY OF THE CAR. THE SUSPENSION CAN MOVE UP OR DOWN AS THE CAR GOES OVER BUMPS TO HELP KEEP THE TIRES ON THE ROAD.

SOLID AXLE VS. INDEPENDENT SUSPENSION

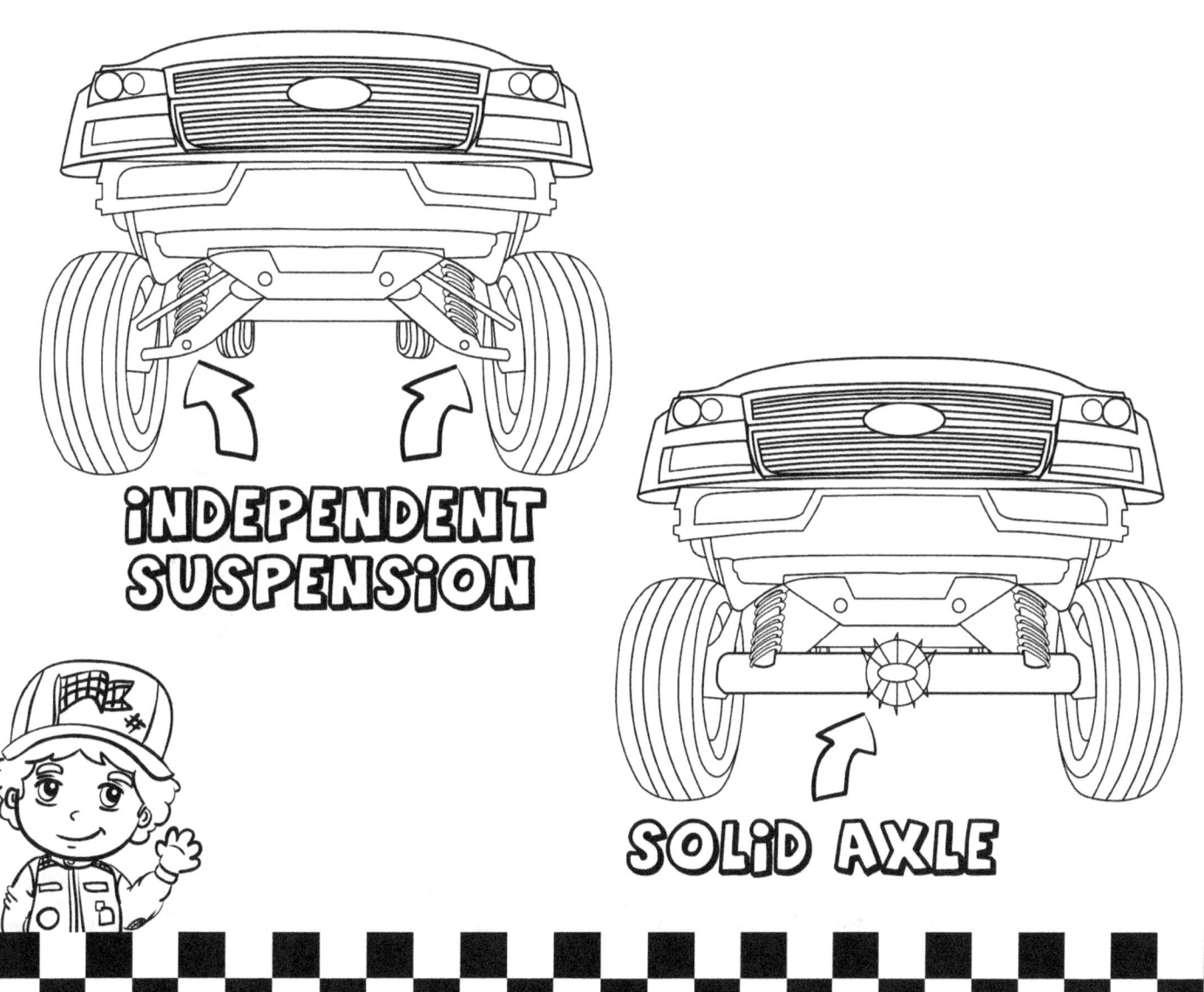

INDEPENDENT SUSPENSION ALLOWS EACH WHEEL TO MOVE UP AND DOWN ON ITS OWN (OR INDEPENDENTLY). A SOLID AXLE IS LIKE A STRAIGHT LINE THAT CONNECTS BOTH WHEELS SO THE WHEELS MOVE UP AND DOWN TOGETHER.

STEERING

THE STEERING WHEEL IS CONNECTED TO THE FRONT 2 WHEELS; TURNING THE STEERING WHEEL ALLOWS THE DRIVER TO CONTROL WHERE THE VEHICLE GOES. A MONSTER TRUCK HAS REAR STEERING, SO THE REAR WHEELS CAN TURN TOO!

TIRES

TIRES ARE THE BLACK RUBBER PART ON THE OUTSIDE OF THE WHEEL. TIRES ARE VERY IMPORTANT. A TIRE IS THE ONLY PART OF THE CAR THAT TOUCHES THE GROUND. TIRES HAVE SPECIAL GROOVES FOR WATER OR DIRT SO THAT THE WHEEL CAN TOUCH THE ROAD EVEN ON RAINY DAYS.

WHEELS

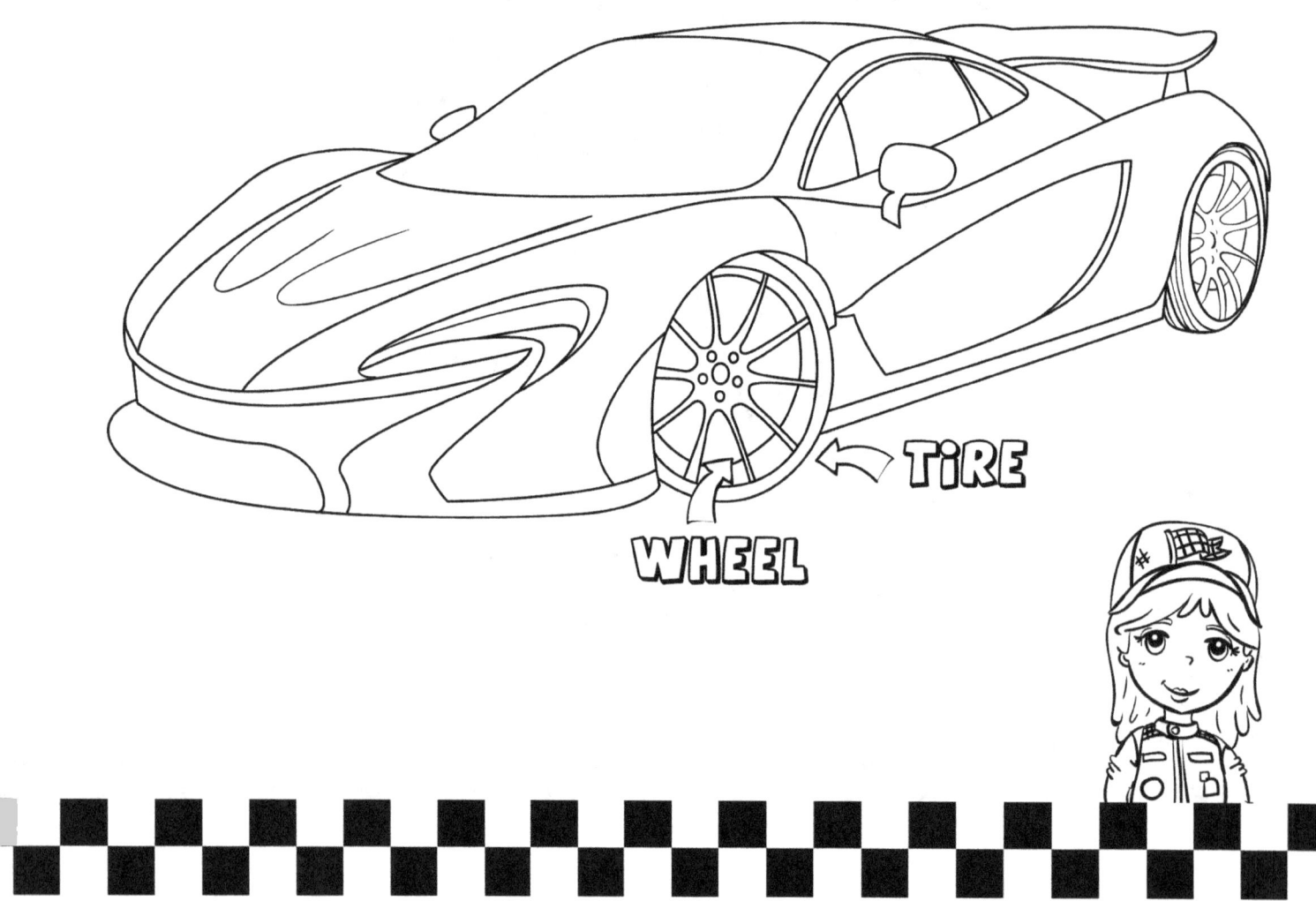

THE WHEEL IS THE METAL PART THAT CONNECTS TO THE CAR. WHEELS ARE USUALLY SILVER BUT CAN BE ANY COLOR.

BRAKES

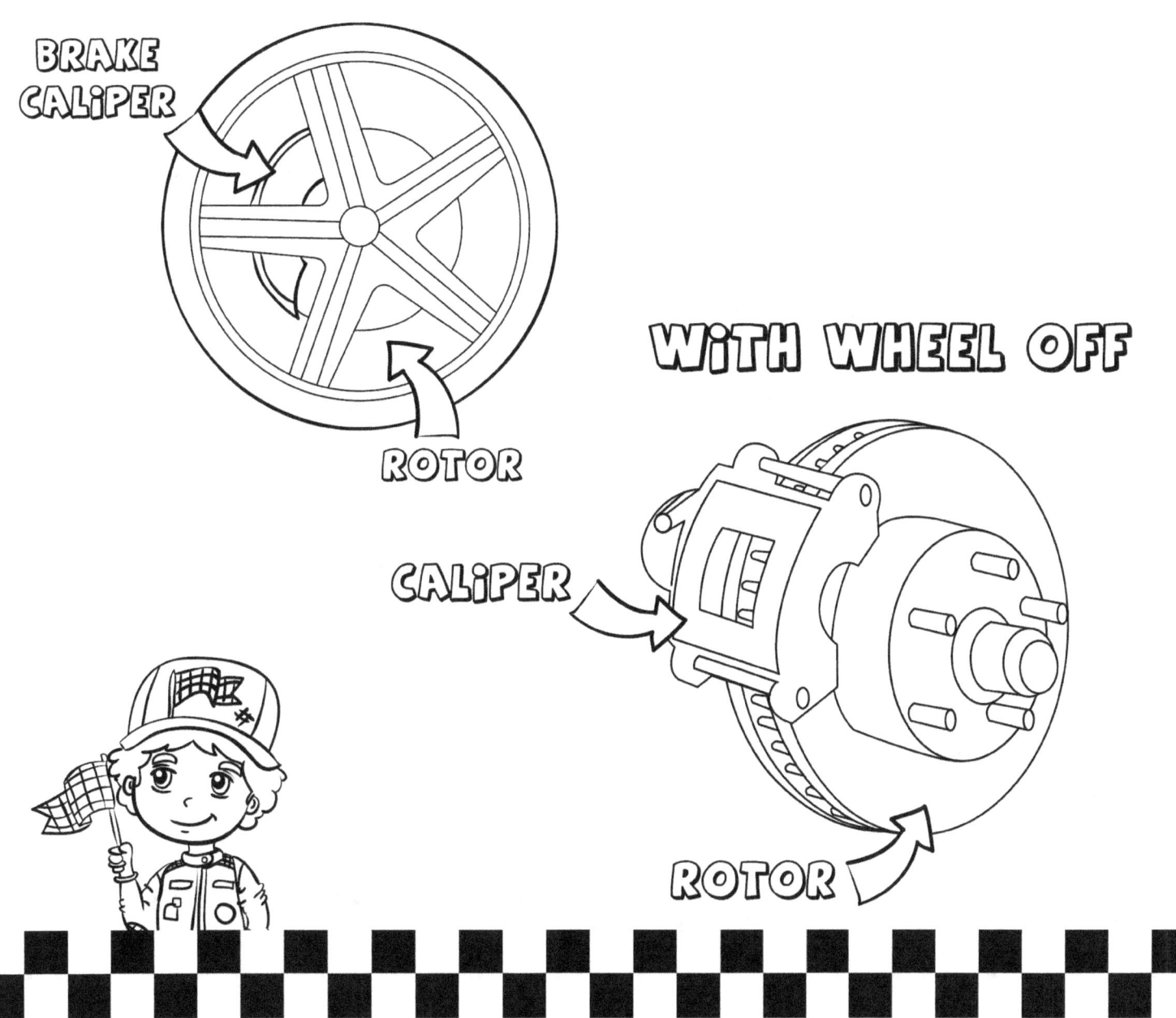

THE BRAKES STOP THE CAR. PUSHING THE BRAKE PEDAL MAKES THE BRAKE CALIPERS SQUEEZE THE ROTORS TO SLOW THE CAR DOWN. BRAKES ON A BIKE WORK THE SAME WAY.

LIGHTS

- HEADLIGHTS IN THE FRONT OF THE CAR LIGHT UP THE ROAD ON DARK NIGHTS.
- TAIL LIGHTS HELP TO MAKE SURE OTHER PEOPLE SEE THE CAR.
- BRAKE LIGHTS WARN OTHERS THAT THE CAR IS SLOWING DOWN, SO THEY'LL NEED TO SLOW DOWN TOO.

POWER WINDOWS

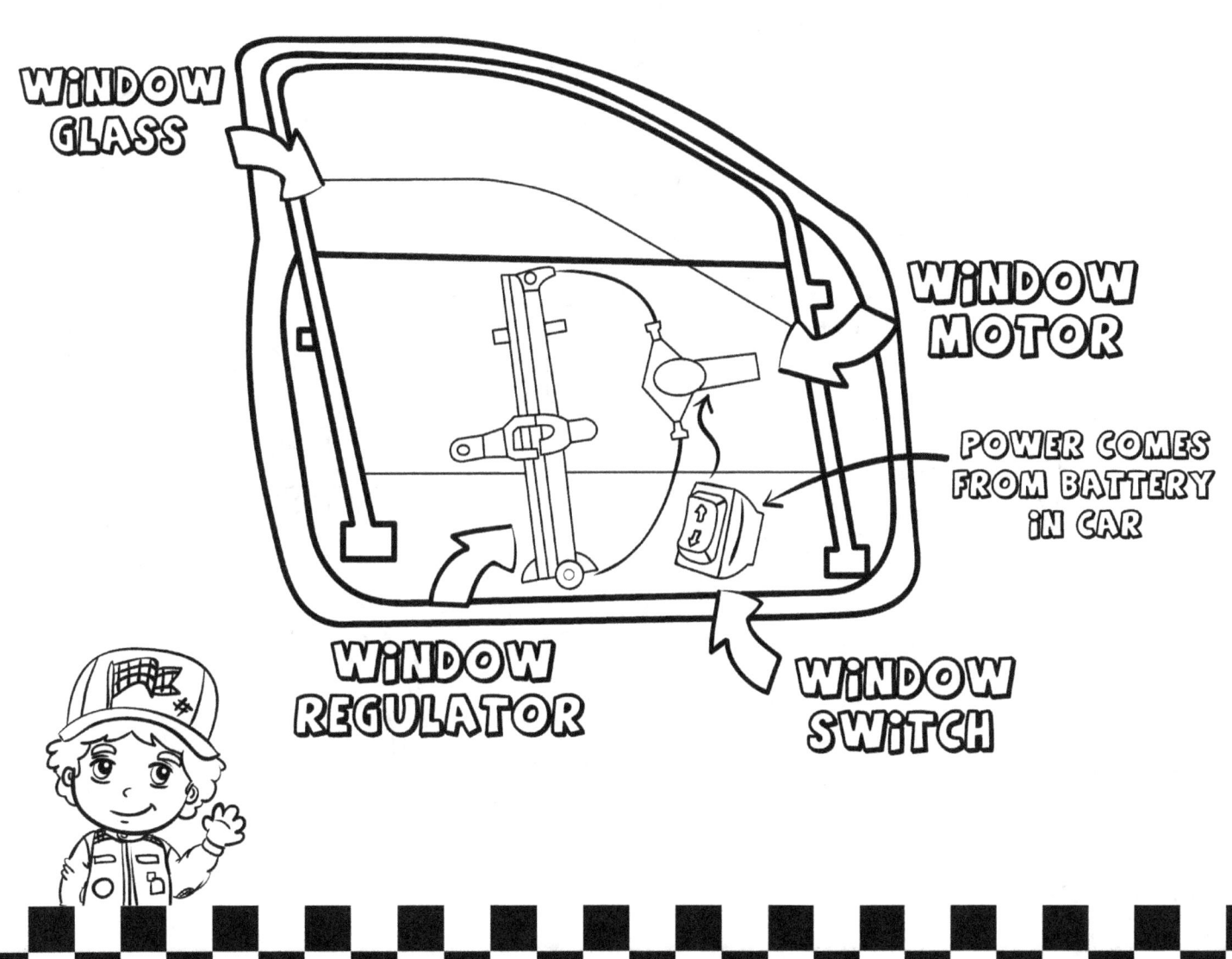

THERE IS A MOTOR IN THE DOOR THAT MOVES THE WINDOWS UP AND DOWN. THE WINDOW SWITCH SENDS ELECTRICITY FROM THE BATTERY TO THE MOTOR WHICH MAKES THE MOTOR MOVE THE WINDOW.

WINDSHIELD WIPERS

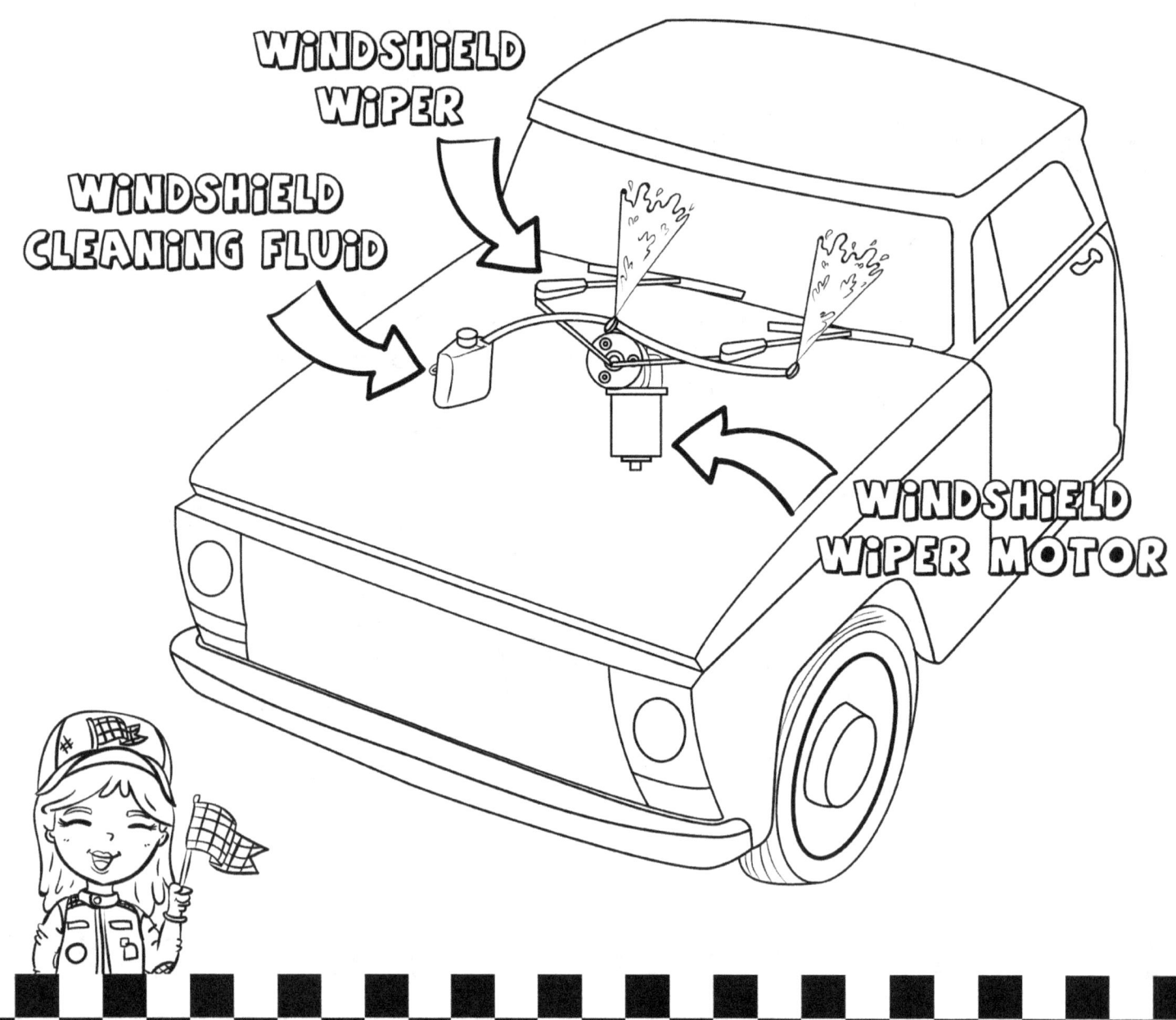

WINDSHIELD WIPERS CLEAN OFF THE WINDSHIELD
ON RAINY DAYS.
THERE IS A MOTOR THAT SPINS THE WIPERS.
THERE IS ALSO A TANK OF CLEANING FLUID IN THE CAR
THAT WILL SQUIRT ONTO THE WINDSHIELD
WHEN THE DRIVER PUSHES A BUTTON.

INSIDE THE CAR

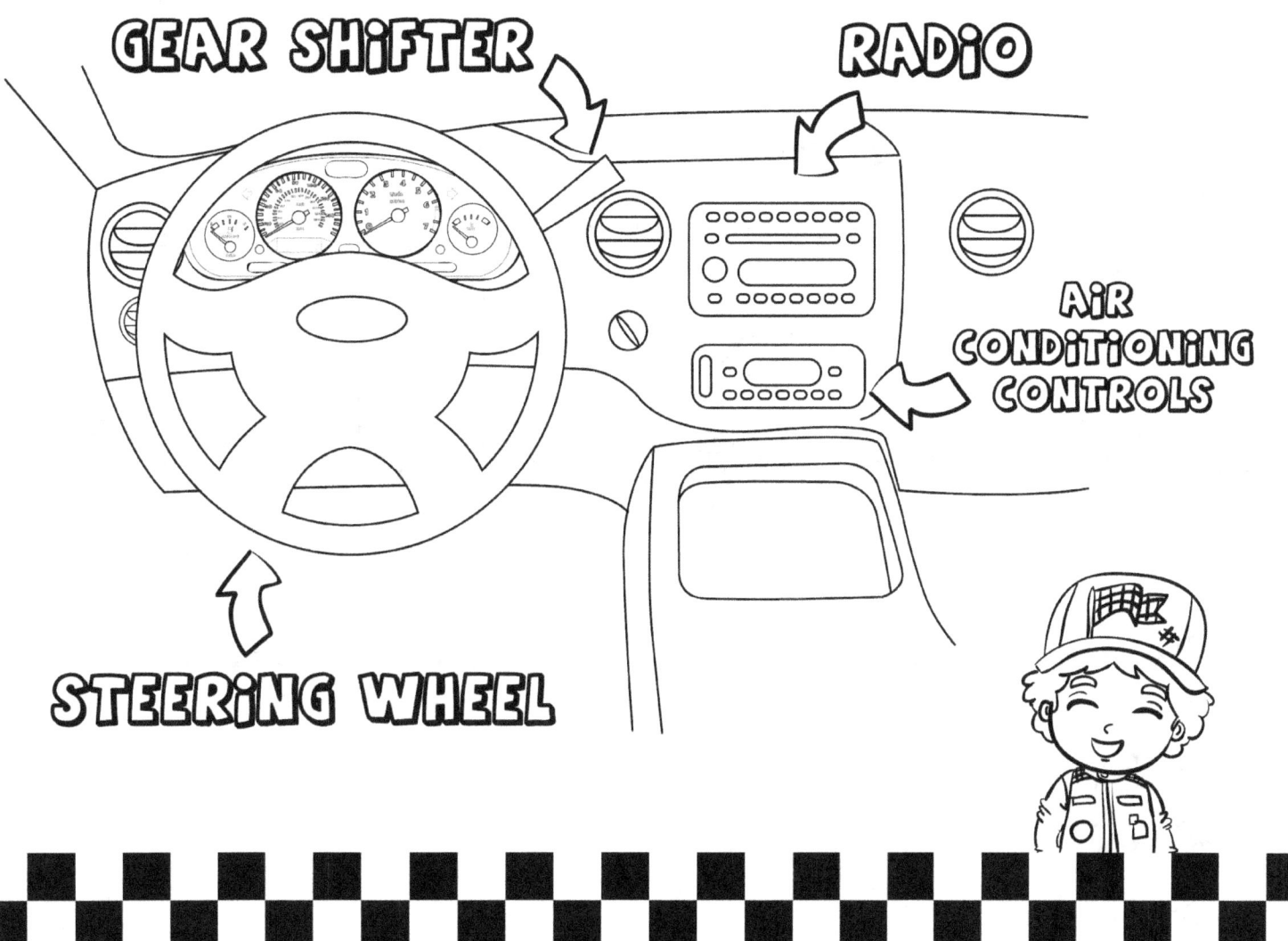

THE DRIVER CAN CONTROL MANY THINGS FROM THE DRIVER SEAT.

- THE STEERING WHEEL CONTROLS WHERE THE CAR GOES.
- THE GEAR SHIFT CONTROLS WHICH GEAR THE TRANSMISSION IS IN.
- THE RADIO CONTROLS THE SOUNDS COMING OUT OF THE SPEAKERS.
- THE AIR CONDITIONING CONTROLS DETERMINE HOW HOT OR COLD IT IS INSIDE THE CAR.

The Little Engineer Coloring Book: Cars and Trucks

INSTRUMENT CLUSTER

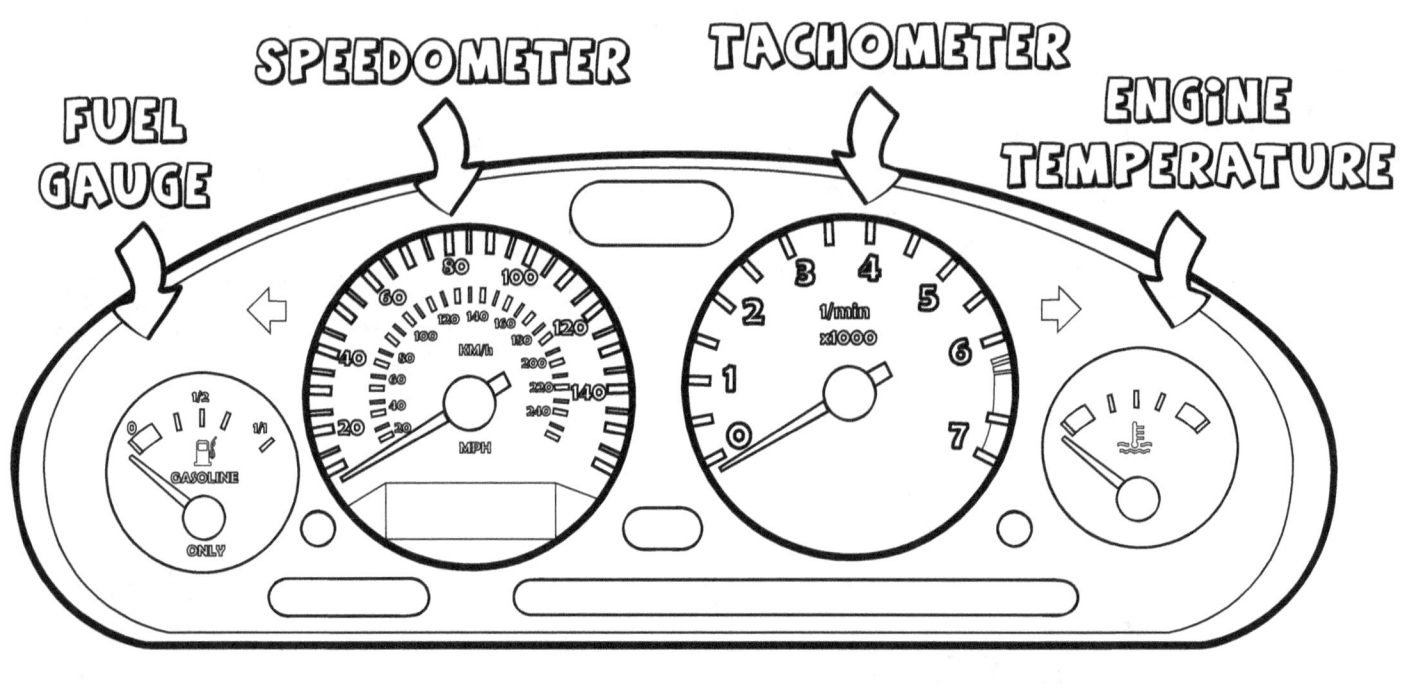

- FUEL GAUGE: SHOWS HOW MUCH FUEL IS IN THE TANK.
- SPEEDOMETER: SHOWS HOW FAST THE CAR IS GOING.
- TACHOMETER: SHOWS HOW FAST THE ENGINE IS SPINNING.
- ENGINE TEMPERATURE: SHOWS THE TEMPERATURE OF THE ENGINE.

HOW DO CARS GET FUEL?

CARS FILL UP THEIR GAS TANKS AT THE GAS STATION.
GAS STATIONS HAVE HUGE TANKS UNDERGROUND
THAT HOLD ALL THEIR FUEL.
TANKER TRUCKS DELIVER FUEL TO THE GAS STATION TANKS.

The Little Engineer Coloring Book: Cars and Trucks

ELECTRIC CAR RECHARGE

AN ELECTRIC CAR DOESN'T HAVE A GAS TANK. IT HAS A BIG BATTERY AND THE BATTERY IS CHARGED WITH A PLUG IN THE GARAGE. SOME PARKING LOTS ALSO HAVE CHARGING STATIONS SO ELECTRIC CARS CAN CHARGE UP WHILE PARKED AT A STORE.

Congratulations! Training Complete

You are now certified and have successfully completed your Cars and Trucks Training!

Go to: thelittleengineerbooks.com/tle-cars-certificate or scan the QR code, and enter your email to get a ready-to-print certificate

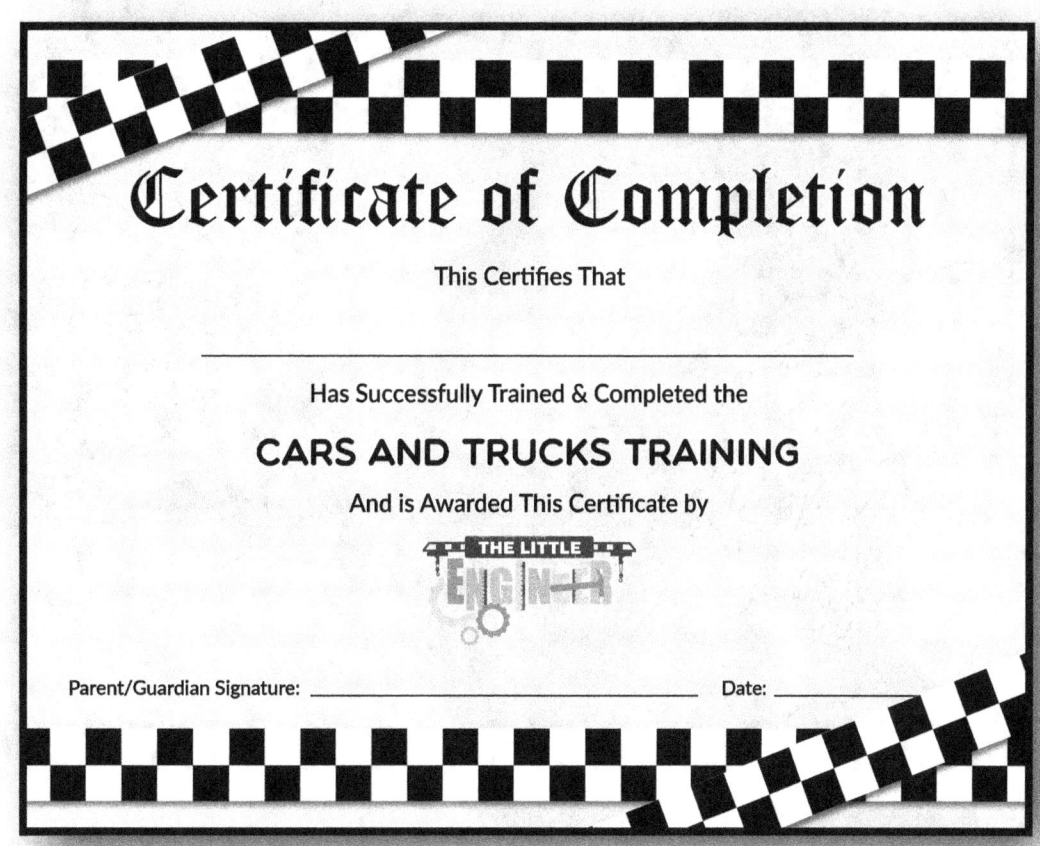

SPECIAL PREVIEW

Check out this short preview of another fun coloring book!

MODERN ROCKETS

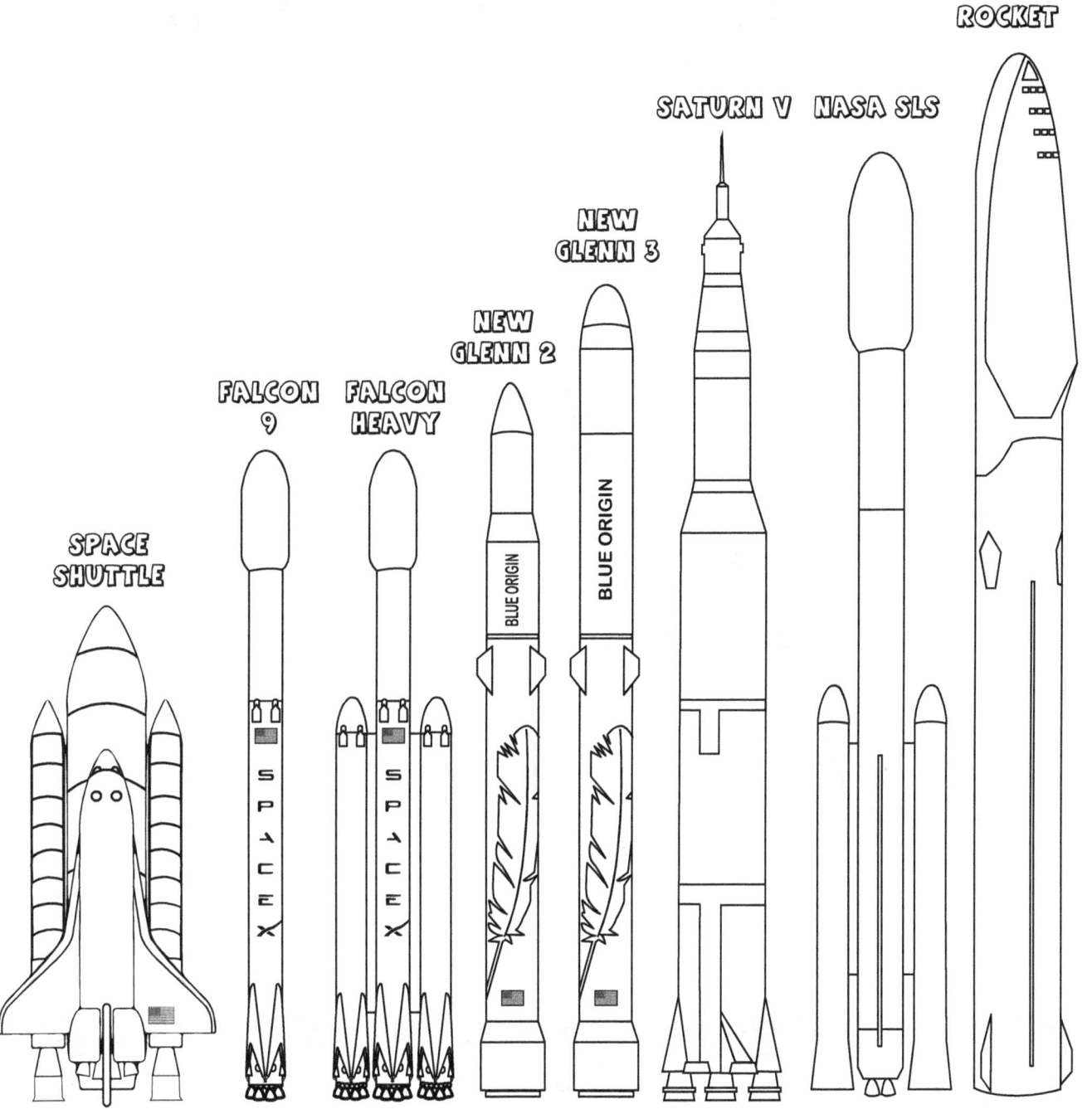

THERE ARE CURRENTLY MANY NEW ROCKETS JUST BECOMING AVAILABLE. THE SATURN V AND SPACE SHUTTLE ARE INCLUDED JUST FOR COMPARISON.

The Little Engineer Coloring Book: Space & Rockets

SPACEX
HOW THE ROCKET LANDS

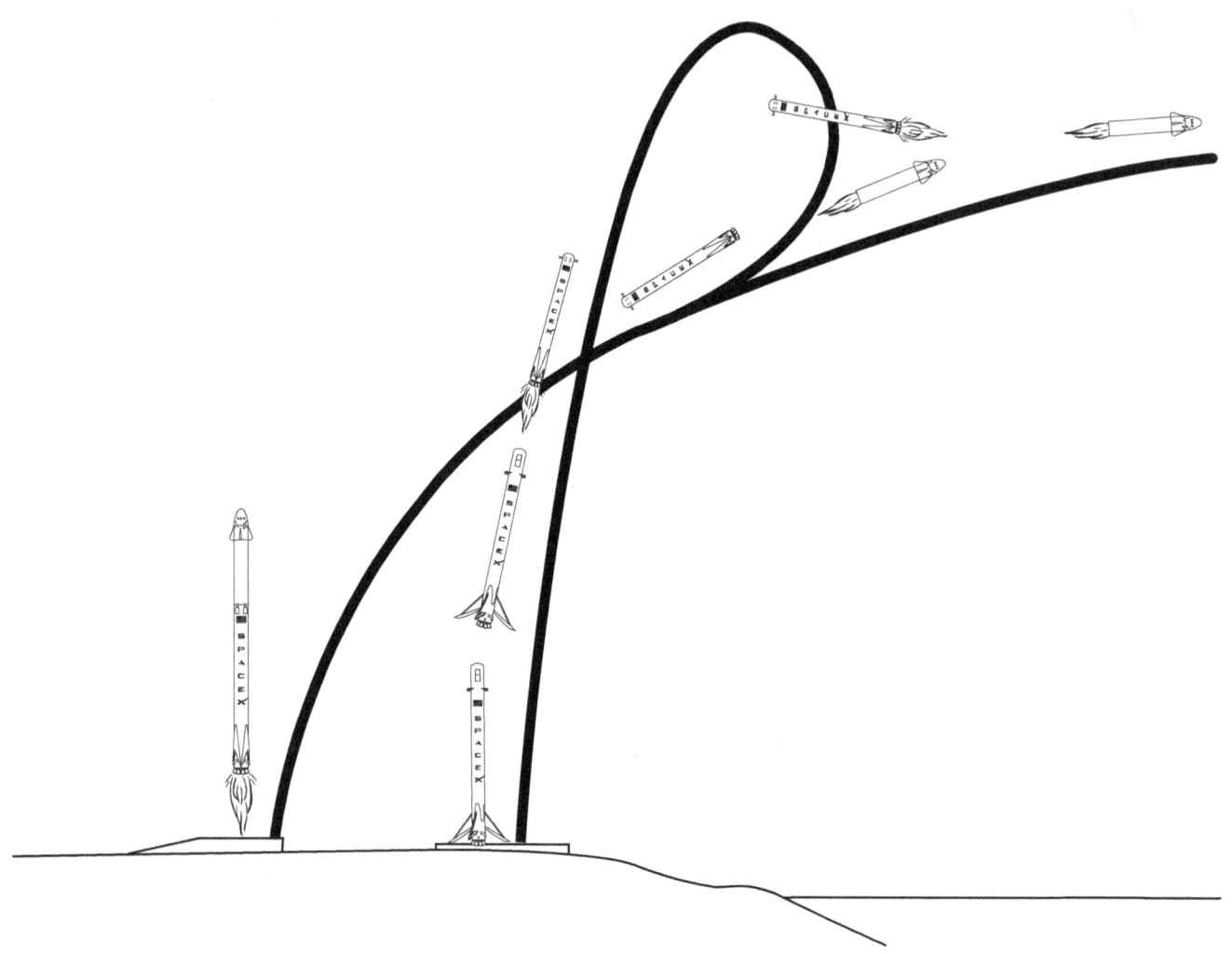

ONCE THE 1ST STAGE OF THE ROCKET SEPARATES FROM THE 2ND STAGE, THE 2ND STAGE WILL CONTINUE TOWARDS ORBIT, BUT THE 1ST STAGE THEN TURNS AROUND AND FLIES BACK TOWARDS A LANDING SPOT ON EARTH.

LAUNCH STAGES

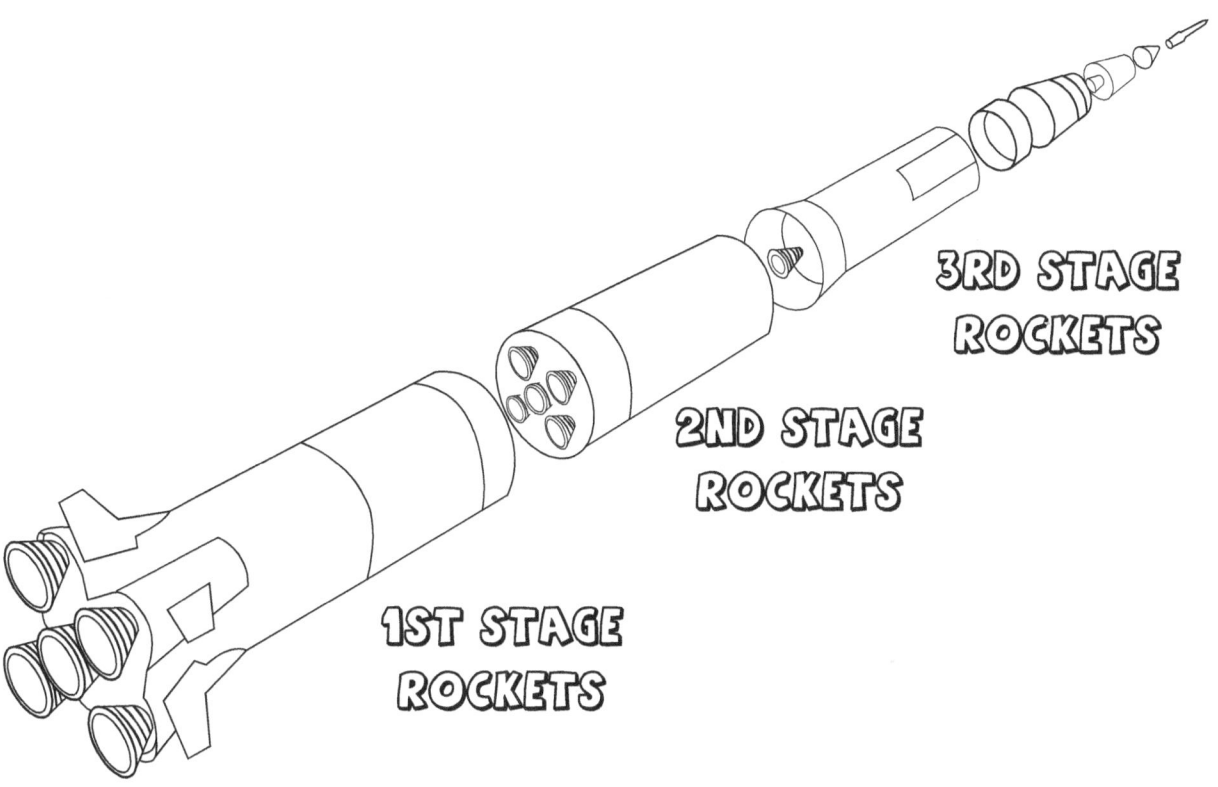

THE ROCKET HAS SEVERAL DIFFERENT SECTIONS CALLED STAGES.

GOING TO THE MOON
STEP 3

STEP 3: FIRE STAGE 3 OF THE ROCKET TO LEAVE EARTH'S ORBIT AND HEAD TOWARD THE MOON.

We hope you enjoyed the book!
Contact us anytime at CreativeIdeasPublishing.com

We are a US based publisher that consist of parents and teachers. We try our best to make products that our kids will love and we hope your kids love them too!

Ask your bookstore for more great titles from
Creative Ideas Publishing!

 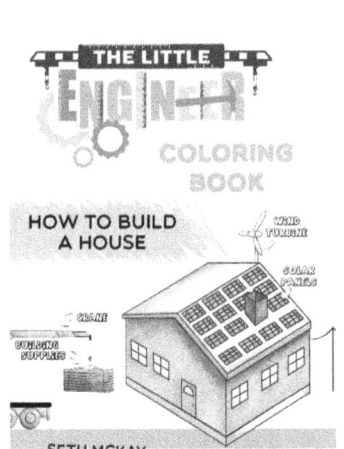  

www.ingramcontent.com/pod-product-compliance
Lightning Source LLC
Chambersburg PA
CBHW081757100526
44592CB00015B/2473